Plastic Analysis and Design of Steel Structures

Plastic Analysis and Design of Steel Structures

M. Bill Wong
Department of Civil Engineering
Monash University, Australia

ELSEVIER

AMSTERDAM • BOSTON • HEIDELBERG • LONDON
NEW YORK • OXFORD • PARIS • SAN DIEGO
SAN FRANCISCO • SINGAPORE • SYDNEY • TOKYO

Butterworth-Heinemann is an imprint of Elsevier

Butterworth-Heinemann is an imprint of Elsevier
30 Corporate Drive, Suite 400, Burlington, MA 01803, USA
Linacre House, Jordan Hill, Oxford OX2 8DP, UK

Library of Congress Cataloging-in-Publication Data
Wong, Bill.
Plastic analysis and design of steel structures/by Bill Wong. -- 1st ed.
p. cm.
Includes bibliographical references and index.
ISBN 978-0-7506-8298-5 (alk. paper)
1. Building, Iron and steel. 2. Structural design. 3. Plastic analysis (Engineering) I. Title.
TA684.W66 2009
624.1'821--dc22

2008027081

British Library Cataloguing-in-Publication Data
A catalogue record for this book is available from the British Library.

ISBN: 978-0-7506-8298-5

For information on all Butterworth–Heinemann publications visit our Web site at www.elsevierdirect.com

Printed in the United States of America
08 09 10 11 12 10 9 8 7 6 5 4 3 2 1

Contents

Preface **ix**

1. Structural Analysis—Stiffness Method **1**
1.1 Introduction 1
1.2 Degrees of Freedom and Indeterminacy 3
1.3 Statically Indeterminate Structures—Direct Stiffness Method 6
1.4 Member Stiffness Matrix 9
1.5 Coordinates Transformation 11
1.6 Member Stiffness Matrix in Global Coordinate System 13
1.7 Assembly of Structure Stiffness Matrix 14
1.8 Load Vector 18
1.9 Methods of Solution 19
1.10 Calculation of Member Forces 20
1.11 Treatment of Internal Loads 27
1.12 Treatment of Pins 32
1.13 Temperature Effects 45
Problems 50
Bibliography 53

2. Plastic Behavior of Structures **55**
2.1 Introduction 55
2.2 Elastic and Plastic Behavior of Steel 55
2.3 Moment–Curvature Relationship in an Elastic–Plastic Range 59
2.4 Plastic Hinge 72
2.5 Plastic Design Concept 73
2.6 Comparison of Linear Elastic and Plastic Designs 73
2.7 Limit States Design 74
2.8 Overview of Design Codes for Plastic Design 75
2.9 Limitations of Plastic Design Method 76
Problems 78
Bibliography 80

3. Plastic Flow Rule and Elastoplastic Analysis **81**
3.1 General Elastoplastic Analysis of Structures 81
3.2 Reduced Plastic Moment Capacity Due to Force Interaction 83
3.3 Concept of Yield Surface 87
3.4 Yield Surface and Plastic Flow Rule 89
3.5 Derivation of General Elastoplastic Stiffness Matrices 92
3.6 Elastoplastic Stiffness Matrices for Sections 95
3.7 Stiffness Matrix and Elastoplastic Analysis 99
3.8 Modified End Actions 101
3.9 Linearized Yield Surface 104
Problems 105
Bibliography 106

4. Incremental Elastoplastic Analysis—Hinge by Hinge Method **107**
4.1 Introduction 107
4.2 Use of Computers for Elastoplastic Analysis 108
4.3 Use of Spreadsheet for Automated Analysis 112
4.4 Calculation of Design Actions and Deflections 116
4.5 Effect of Force Interaction on Plastic Collapse 119
4.6 Plastic Hinge Unloading 124
4.7 Distributed Loads in Elastoplastic Analysis 126
Problems 130
Bibliography 137

5. Manual Methods of Plastic Analysis **139**
5.1 Introduction 139
5.2 Theorems of Plasticity 139
5.3 Mechanism Method 141
5.4 Statical Method 142
5.5 Uniformly Distributed Loads (UDL) 144
5.6 Continuous Beams and Frames 147
5.7 Calculation of Member Forces at Collapse 155
5.8 Effect of Axial Force on Plastic Collapse Load 157
Problems 159
Bibliography 162

6. Limit Analysis by Linear Programming **163**
6.1 Introduction 163
6.2 Limit Analysis Theorems as Constrained Optimization Problems 164
6.3 Spreadsheet Solution of Simple Limit Analysis Problems 166

6.4 General Description of the Discrete Plane Frame Problem 175
6.5 A Simple MATLAB Implementation for Static Limit Analysis 183
6.6 A Note on Optimal Plastic Design of Frames 189
Bibliography 193

7. Factors Affecting Plastic Collapse 195
7.1 Introduction 195
7.2 Plastic Rotation Capacity 195
7.3 Effect of Settlement 200
7.4 Effect of High Temperature 206
7.5 Second-Order Effects 213
Problems 215
Bibliography 216

8. Design Consideration 219
8.1 Introduction 219
8.2 Serviceability Limit State Requirements 219
8.3 Ultimate Limit State Requirements 223
Bibliography 231

Answers 233

Index 237

Preface

The plastic method has been used extensively by engineers for the design of steel structures, including simple beams, continuous beams, and simple portal frames. Traditionally, the analysis is based on the rigid-plastic theory whereby the plastic collapse load is evaluated through virtual work formulation in which elastic deflection is ignored. For more complex frames, specialist computer packages for elastoplastic analysis are usually employed. Current publications on plastic design method provide means of analysis based on either virtual work formulation or sophisticated plastic theory contained in specialist computer packages. This book aims to bridge this gap.

The advent of computers has enabled practicing engineers to perform linear and nonlinear elastic analysis on a daily basis using computer programs widely available commercially. The results from computer analysis are transferred routinely to tools with automated calculation formats such as spreadsheets for design. The use of this routine procedure is commonplace for design based on elastic, geometrically nonlinear analysis. However, commercially available computer programs for plastic analysis are still a rarity among the engineering community.

This book emphasizes a plastic analysis method based on the hinge by hinge concept. Frames of any degree of complexity can be analyzed plastically using this method. This method is based on the elastoplastic analysis procedure where a linear elastic analysis, performed either manually or by computers, is used between the formation of consecutive plastic hinges. The results of the linear elastic analysis are used in a proforma created in a spreadsheet environment where the next plastic hinge formation can be predicted automatically and the corresponding culmulative forces and deflections calculated. In addition, a successive approximation method is described to take account of the effect of force interaction on the evaluation of the collapse load of a structure. This method can be performed using results from analysis obtained from most commercially available computer programs.

The successive approximation method is an indirect way to obtain the collapse load of structures using iterative procedures. For direct calculation of the collapse load without using iterative procedures, special formulations, possibly with ad-hoc computer programming, according to the plastic theory must be used. Nowadays, the stiffness method is the most popular and recognized method for structural analysis. This book provides a theoretical treatment for derivation of the stiffness matrices for different states of plasticity in an element for the stiffness method of analysis. The theory is based on the plastic flow rule and the concept of yield surface is introduced.

An introduction to the use of the linear programming technique for plastic analysis is provided in a single chapter in this book. This powerful and advanced method for plastic analysis is described in detail using optimization procedures. Its use is important in an automated computational environment and is particularly important for researchers working in the area of nonlinear structural plastic analysis. This chapter was written by Professor Francis Tin-Loi, a prominent researcher in the use of mathematical programming methods for plastic analysis of structures.

In this book, new insights into various issues related to plastic analysis and design are given, such as the effect of high temperature on plastic collapse load and the use of plastic rotation capacity as a limit state for plastic design. Based on the elastoplastic approach, an interpolation procedure is introduced to calculate the design forces and deflections at the design load level rather than at the collapse load level.

In the final chapter of this book, a comparison among design codes from Australia, Europe, and the United States for plastic design method is given. This comparison enables practicing engineers to understand the issues involved in the plastic design procedures and the limitations imposed by this design method.

Bill Wong

CHAPTER 1

Structural Analysis—Stiffness Method

1.1 Introduction

Computer programs for plastic analysis of framed structures have been in existence for some time. Some programs, such as those developed earlier by, among others, Wang,[1] Jennings and Majid,[2] and Davies,[3] and later by Chen and Sohal,[4] perform plastic analysis for frames of considerable size. However, most of these computer programs were written as specialist programs specifically for linear or nonlinear plastic analysis. Unlike linear elastic analysis computer programs, which are commonly available commercially, computer programs for plastic analysis are not as accessible. Indeed, very few, if any, are being used for daily routine design in engineering offices. This may be because of the perception by many engineers that the plastic design method is used only for certain types of usually simple structures, such as beams and portal frames. This perception discourages commercial software developers from developing computer programs for plastic analysis because of their limited applications.

Contrary to the traditional thinking that plastic analysis is performed either by simple manual methods for simple structures or by sophisticated computer programs written for more general applications, this book intends to introduce general plastic analysis methods, which take advantage of the availability of modern computational tools, such as linear elastic analysis programs and spreadsheet applications. These computational tools are in routine use in most engineering design offices nowadays. The powerful number-crunching capability of these tools enables plastic analysis and design to be performed for structures of virtually any size.

The amount of computation required for structural analysis is largely dependent on the degree of statical indeterminacy of the structure. For determinate structures, use of equilibrium conditions alone will enable the reactions and internal forces to be determined. For indeterminate structures, internal forces are calculated by considering both equilibrium and compatibility conditions, through which some methods of structural analysis suitable for computer applications have been developed. The use of these methods for analyzing indeterminate structures is usually not simple, and computers are often used for carrying out these analyses. Most structures in practice are statically indeterminate.

Structural analysis, whether linear or nonlinear, is mostly based on matrix formulations to handle the enormous amount of numerical data and computations. Matrix formulations are suitable for computer implementation and can be applied to two major methods of structural analysis: the flexibility (or force) method and the stiffness (or displacement) method.

The flexibility method is used to solve equilibrium and compatibility equations in which the reactions and member forces are formulated as unknown variables. In this method, the degree of statical indeterminacy needs to be determined first and a number of unknown forces are chosen and released so that the remaining structure, called the *primary structure*, becomes determinate. The primary structure under the externally applied loads is analyzed and its displacement is calculated. A unit value for each of the chosen released forces, called *redundant forces*, is then applied to the primary structure (without the externally applied loads) so that, from the force-displacement relationship, displacements of the structure are calculated. The structure with each of the redundant forces is called the *redundant structure*. The compatibility conditions based on the deformation between the primary structure and the redundant structures are used to set up a matrix equation from which the redundant forces can be solved.

The solution procedure for the force method requires selection of the redundant forces in the original indeterminate structure and the subsequent establishment of the matrix equation from the compatibility conditions. This procedure is not particularly suitable for computer programming and the force method is therefore usually used only for simple structures.

In contrast, formulation of the matrix equations for the stiffness method is done routinely and the solution procedure is systematic. Therefore, the stiffness method is adopted in most structural analysis computer programs. The stiffness method is particularly useful for structures with a high degree of statical indeterminacy, although it can be used for both determinate and indeterminate structures. The stiffness method is used in the elastoplastic analysis described in this book and the basis of this method is given in this chapter.

In particular, the direct stiffness method, a variant of the general stiffness method, is described. For a brief history of the stiffness method, refer to the review by Samuelsson and Zienkiewicz.[5]

1.2 Degrees of Freedom and Indeterminacy

Plastic analysis is used to obtain the behavior of a structure at collapse. As the structure approaches its collapse state when the loads are increasing, the structure becomes increasingly flexible in its stiffness. Its flexibility at any stage of loading is related to the degree of statical indeterminacy, which keeps decreasing as plastic hinges occur with the increasing loads. This section aims to describe a method to distinguish between determinate and indeterminate structures by examining the degrees of freedom of structural frames. The number of degrees of freedom of a structure denotes the independent movements of the structural members at the joints, including the supports. Hence, it is an indication of the size of the structural problem. The degrees of freedom of a structure are counted in relation to a reference coordinate system.

External loads are applied to a structure causing movements at various locations. For frames, these locations are usually defined at the joints for calculation purposes. Thus, the maximum number of independent displacements, including both rotational and translational movements at the joints, is equal to the number of degrees of freedom of the structure. To identify the number of degrees of freedom of a structure, each independent displacement is assigned a number, called the freedom code, in ascending order in the global coordinate system of the structure.

Figure 1.1 shows a frame with 7 degrees of freedom. Note that the pinned joint at C allows the two members BC and CD to rotate independently, thus giving rise to two freedoms in rotation at the joint.

In structural analysis, the degree of statical indeterminacy is important, as its value may determine whether the structure

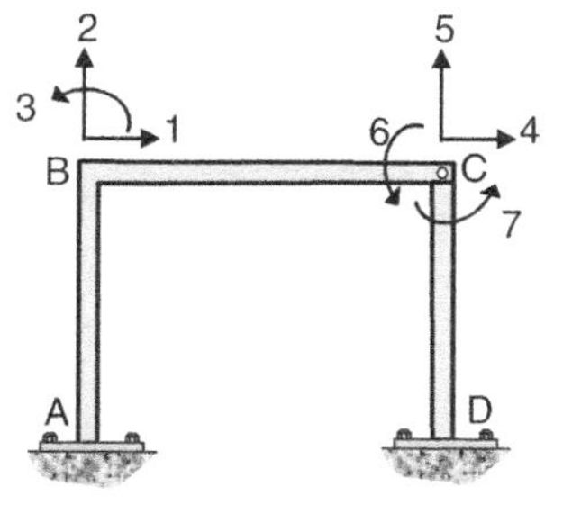

FIGURE 1.1. Degrees of freedom of a frame.

is globally unstable or stable. If the structure is stable, the degree of statical indeterminacy is, in general, proportional to the level of complexity for solving the structural problem.

The method described here for determining the degree of statical indeterminacy of a structure is based on that by Rangasami and Mallick.[6]

Only plane frames will be dealt with here, although the method can be extended to three-dimensional frames.

1.2.1 Degree of Statical Indeterminacy of Frames

For a free member in a plane frame, the number of possible displacements is three: horizontal, vertical, and rotational. If there are n members in the structure, the total number of possible displacements, denoted by m, before any displacement restraints are considered, is

$$m = 3n \tag{1.1}$$

For two members connected at a joint, some or all of the displacements at the joint are common to the two members and these common displacements are considered restraints. In this method for determining the degree of statical indeterminacy, every joint is considered as imposing r number of restraints if the number of common displacements between the members is r. The ground or foundation is considered as a noncounting member and has no freedom. Figure 1.2 indicates the value of r for each type of joints or supports in a plane frame.

For pinned joints with multiple members, the number of pinned joints, p, is counted according to Figure 1.3. For example, for a four-member pinned connection shown in Figure 1.3, a first joint is counted by considering the connection of two members, a second joint by the third member, and so on. The total number of pinned joints for a four-member connection is therefore equal to three. In general, the number of pinned joints connecting n members is $p = n - 1$. The same method applies to fixed joints.

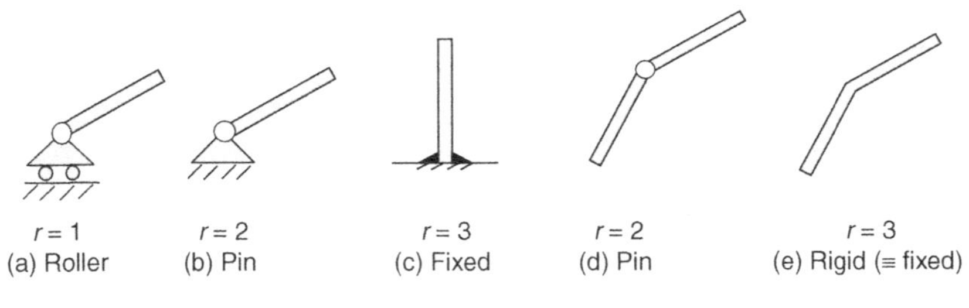

FIGURE 1.2. Restraints of joints.

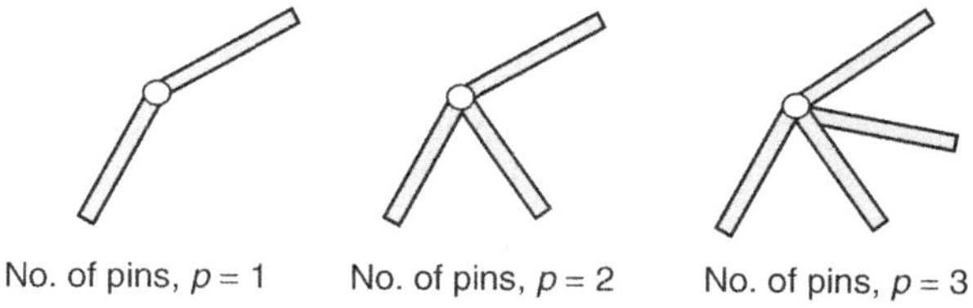

FIGURE 1.3. Method for joint counting.

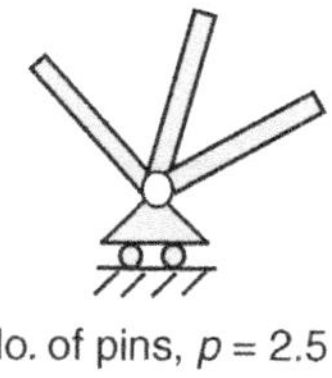

FIGURE 1.4. Joint counting of a pin with roller support.

For a connection at a roller support, as in the example shown in Figure 1.4, it can be calculated that $p = 2.5$ pinned joints and that the total number of restraints is $r = 5$.

The degree of statical indeterminacy, f_r, of a structure is determined by

$$f_r = m - \sum r \tag{1.2}$$

a. If $f_r = 0$, the frame is stable and statically determinate.
b. If $f_r < 0$, the frame is stable and statically indeterminate to the degree f_r.
c. If $f_r > 0$, the frame is unstable.

Note that this method does not examine external instability or partial collapse of the structure.

Example 1.1 Determine the degree of statical indeterminacy for the pin-jointed truss shown in Figure 1.5.

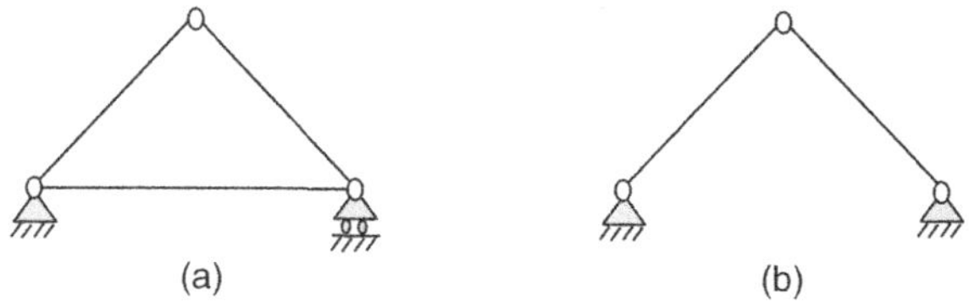

FIGURE 1.5. Determination of degree of statical indeterminacy in Example 1.1.

Solution. For the truss in Figure 1.5a, number of members $n = 3$; number of pinned joints $p = 4.5$.

Hence, $f_r = 3 \times 3 - 2 \times 4.5 = 0$ and the truss is a determinate structure. For the truss in Figure 1.5b, number of members $n = 2$; number of pinned joints $p = 3$.

Hence, $f_r = 3 \times 2 - 2 \times 3 = 0$ and the truss is a determinate structure.

Example 1.2 Determine the degree of statical indeterminacy for the frame with mixed pin and rigid joints shown in Figure 1.6.

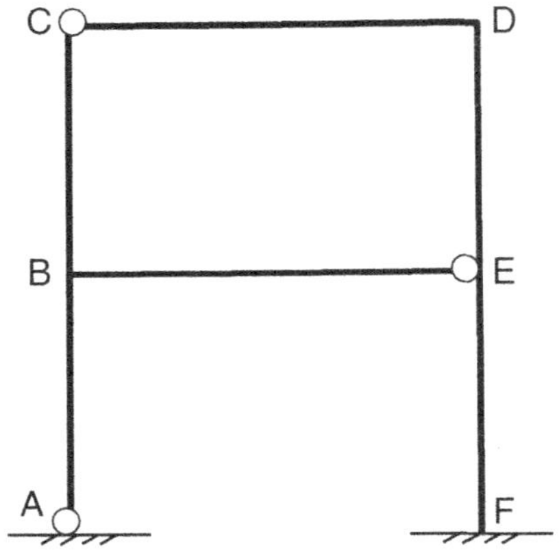

FIGURE 1.6. Determination of degree of statical indeterminacy in Example 1.2.

Solution. For this frame, a member is counted as one between two adjacent joints. Number of members = 6; number of rigid (or fixed) joints = 5. Note that the joint between DE and EF is a rigid one, whereas the joint between BE and DEF is a pinned one. Number of pinned joints = 3.

Hence, $f_r = 3 \times 6 - 3 \times 5 - 2 \times 3 = -3$ and the frame is an indeterminate structure to the degree 3.

1.3 Statically Indeterminate Structures—Direct Stiffness Method

The spring system shown in Figure 1.7 demonstrates the use of the stiffness method in its simplest form. The single degree of freedom structure consists of an object supported by a linear spring obeying Hooke's law. For structural analysis, the weight, F, of the object and the spring constant (or stiffness), K, are usually known. The purpose

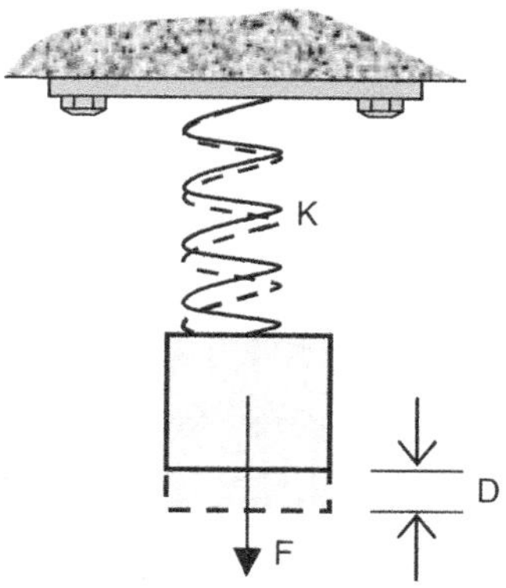

FIGURE 1.7. Load supported by linear spring.

of the structural analysis is to find the vertical displacement, D, and the internal force in the spring, P.

From Hooke's law,

$$F = KD \tag{1.3}$$

Equation (1.3) is in fact the equilibrium equation of the system. Hence, the displacement, D, of the object can be obtained by

$$D = F/K \tag{1.4}$$

The displacement, d, of the spring is obviously equal to D. That is,

$$d = D \tag{1.5}$$

The internal force in the spring, P, can be found by

$$P = Kd \tag{1.6}$$

In this simple example, the procedure for using the stiffness method is demonstrated through Equations (1.3) to (1.6). For a structure composed of a number of structural members with n degrees of freedom, the equilibrium of the structure can be described by a number of equations analogous to Equation (1.3). These equations can be expressed in matrix form as

$$\{F\}_{n\times 1} = [K]_{n\times n}\{D\}_{n\times 1} \tag{1.7}$$

where $\{F\}_{n\times 1}$ is the load vector of size $(n \times 1)$ containing the external loads, $[K]_{n\times n}$ is the structure stiffness matrix of size $(n \times n)$ corresponding to the spring constant K in a single degree system shown in Figure 1.7, and $\{D\}_{n\times 1}$ is the displacement vector of size $(n \times 1)$ containing the unknown displacements at designated locations, usually at the joints of the structure.

The unknown displacement vector can be found by solving Equation (1.7) as

$$\{D\} = [K]^{-1}\{F\} \tag{1.8}$$

Details of the formation of $\{F\}$, $[K]$, and $\{D\}$ are given in the following sections.

1.3.1 Local and Global Coordinate Systems

A framed structure consists of discrete members connected at joints, which may be pinned or rigid. In a local coordinate system for a member connecting two joints i and j, the member forces and the corresponding displacements are shown in Figure 1.8, where the axial forces are acting along the longitudinal axis of the member and the shear forces are acting perpendicular to its longitudinal axis.

In Figure 1.8, $M_{i,j}$, $\theta_{i,j}$ = bending moments and corresponding rotations at ends i, j, respectively; $N_{i,j}$, $u_{i,j}$ are axial forces and corresponding axial deformations at ends i, j, respectively; and $Q_{i,j}$, $v_{i,j}$ are shear forces and corresponding transverse displacements at ends i, j, respectively. The directions of the actions and movements shown in Figure 1.8 are positive when using the stiffness method.

As mentioned in Section 1.2, the freedom codes of a structure are assigned in its global coordinate system. An example of a member forming part of the structure with a set of freedom codes (1, 2, 3, 4, 5, 6) at its ends is shown in Figure 1.9. At either end of the member, the direction in which the member is restrained from movement is assigned a freedom code "zero," otherwise a nonzero freedom code is assigned. The relationship for forces and displacements between local and global coordinate systems will be established in later sections.

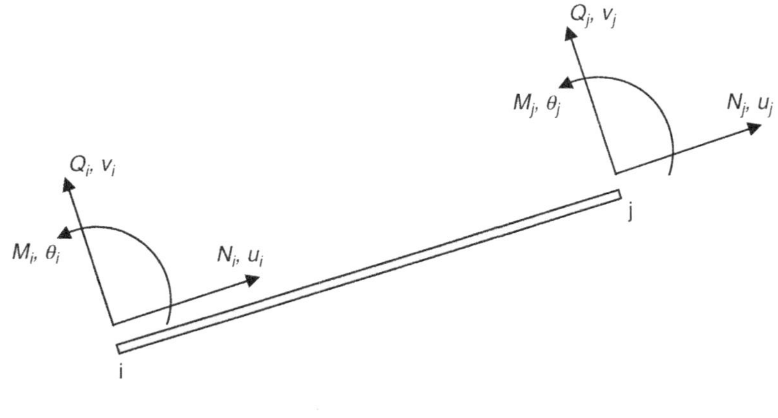

FIGURE 1.8. Local coordinate system for member forces and displacements.

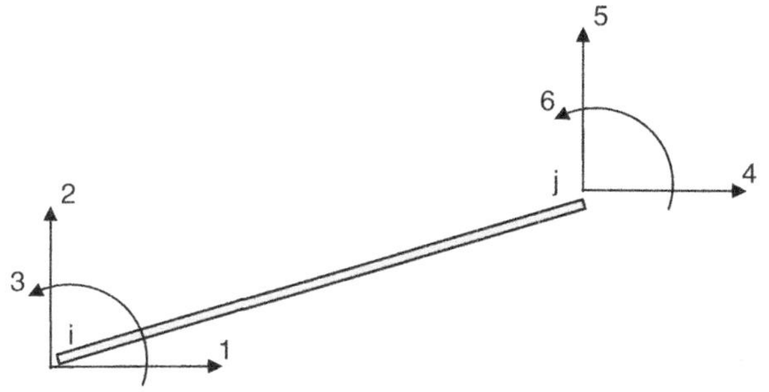

FIGURE 1.9. Freedom codes of a member in a global coordinate system.

1.4 Member Stiffness Matrix

The structure stiffness matrix $[K]$ is assembled on the basis of the equilibrium and compatibility conditions between the members. For a general frame, the equilibrium matrix equation of a member is

$$\{P\} = [K_e]\{d\} \tag{1.9}$$

where $\{P\}$ is the member force vector, $[K_e]$ is the member stiffness matrix, and $\{d\}$ is the member displacement vector, all in the member's local coordinate system. The elements of the matrices in Equation (1.9) are given as

$$\{P\} = \begin{Bmatrix} N_i \\ Q_i \\ M_i \\ N_j \\ Q_j \\ M_j \end{Bmatrix}; [K_e] = \begin{bmatrix} K_{11} & 0 & 0 & K_{14} & 0 & 0 \\ 0 & K_{22} & K_{23} & 0 & K_{25} & K_{26} \\ 0 & K_{32} & K_{33} & 0 & K_{35} & K_{36} \\ K_{41} & 0 & 0 & K_{44} & 0 & 0 \\ 0 & K_{52} & K_{53} & 0 & K_{55} & K_{56} \\ 0 & K_{62} & K_{63} & 0 & K_{65} & K_{66} \end{bmatrix}; \{d\} = \begin{Bmatrix} u_i \\ v_i \\ \theta_i \\ u_j \\ v_j \\ \theta_j \end{Bmatrix}$$

where the elements of $\{P\}$ and $\{d\}$ are shown in Figure 1.8.

1.4.1 Derivation of Elements of Member Stiffness Matrix

A member under axial forces N_i and N_j acting at its ends produces axial displacements u_i and u_j as shown in Figure 1.10. From the stress-strain relation, it can be shown that

$$N_i = \frac{EA}{L}(u_i - u_j) \tag{1.10a}$$

$$N_j = \frac{EA}{L}(u_j - u_i) \tag{1.10b}$$

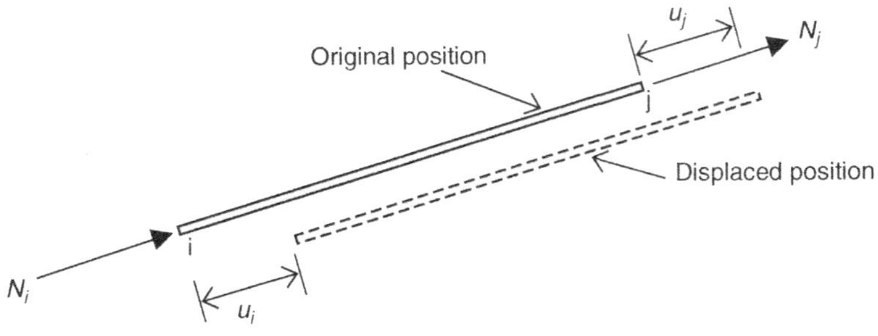

FIGURE 1.10. Member under axial forces.

where E is Young's modulus, A is cross-sectional area, and L is length of the member. Hence, $K_{11} = -K_{14} = -K_{41} = K_{44} = \frac{EA}{L}$.

For a member with shear forces Q_i, Q_j and bending moments M_i, M_j acting at its ends as shown in Figure 1.11, the end displacements and rotations are related to the bending moments by the slope-deflection equations as

$$M_i = \frac{2EI}{L}\left[2\theta_i + \theta_j - \frac{3(v_j - v_i)}{L}\right] \tag{1.11a}$$

$$M_j = \frac{2EI}{L}\left[2\theta_j + \theta_i - \frac{3(v_j - v_i)}{L}\right] \tag{1.11b}$$

Hence, $K_{62} = -K_{65} = \frac{6EI}{L^2}$, $K_{63} = \frac{2EI}{L}$, and $K_{66} = \frac{4EI}{L}$.

By taking the moment about end j of the member in Figure 1.11, we obtain

$$Q_i = \frac{M_i + M_j}{L} = \frac{2EI}{L^2}\left[3\theta_i + 3\theta_j - \frac{6(v_j - v_i)}{L}\right] \tag{1.12a}$$

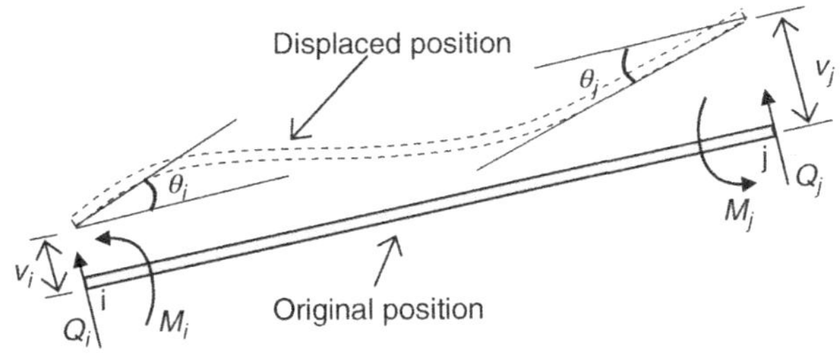

FIGURE 1.11. Member under shear forces and bending moments.

Also, by taking the moment about end i of the member, we obtain

$$Q_j = -\left(\frac{M_i + M_j}{L}\right) = -Q_i \tag{1.12b}$$

Hence,

$$K_{22} = K_{55} = -K_{25} = -K_{52} = \frac{12EI}{L^3} \text{ and } K_{23} = K_{26} = -K_{53} = -K_{66} = \frac{6EI}{L^2}.$$

In summary, the resulting member stiffness matrix is symmetric about the diagonal:

$$[K_e] = \begin{bmatrix} \frac{EA}{L} & 0 & 0 & -\frac{EA}{L} & 0 & 0 \\ 0 & \frac{12EI}{L^3} & \frac{6EI}{L^2} & 0 & -\frac{12EI}{L^3} & \frac{6EI}{L^2} \\ 0 & \frac{6EI}{L^2} & \frac{4EI}{L} & 0 & -\frac{6EI}{L^2} & \frac{2EI}{L} \\ -\frac{EA}{L} & 0 & 0 & \frac{EA}{L} & 0 & 0 \\ 0 & -\frac{12EI}{L^3} & -\frac{6EI}{L^2} & 0 & \frac{12EI}{L^3} & -\frac{6EI}{L^2} \\ 0 & \frac{6EI}{L^2} & \frac{2EI}{L} & 0 & -\frac{6EI}{L^2} & \frac{4EI}{L} \end{bmatrix} \tag{1.13}$$

1.5 Coordinates Transformation

In order to establish the equilibrium conditions between the member forces in the local coordinate system and the externally applied loads in the global coordinate system, the member forces are transformed into the global coordinate system by force resolution. Figure 1.12 shows a member inclined at an angle α to the horizontal.

1.5.1 Load Transformation

The forces in the global coordinate system shown with superscript "g" in Figure 1.12 are related to those in the local coordinate system by

$$H_i^g = N_i \cos\alpha - Q_i \sin\alpha \tag{1.14a}$$

$$V_i^g = N_i \sin\alpha + Q_i \cos\alpha \tag{1.14b}$$

$$M_i^g = M_i \tag{1.14c}$$

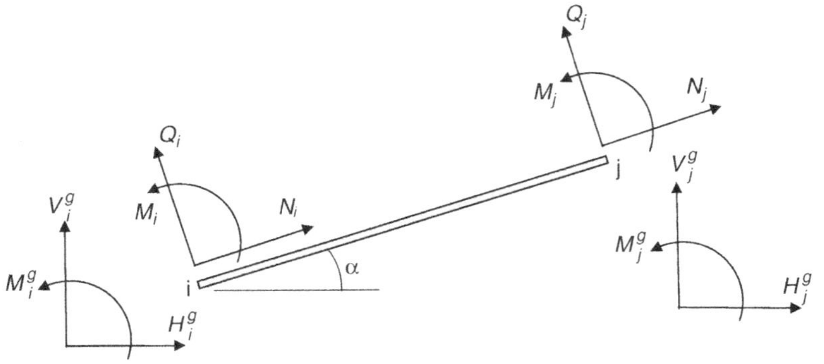

FIGURE 1.12. Forces in the local and global coordinate systems.

Similarly,

$$H_j^g = N_j \cos\alpha - Q_j \sin\alpha \tag{1.14d}$$

$$V_j^g = N_j \sin\alpha + Q_j \cos\alpha \tag{1.14e}$$

$$M_j^g = M_j \tag{1.14f}$$

In matrix form, Equations (1.14a) to (1.14f) can be expressed as

$$\{F_e^g\} = [T]\{P\} \tag{1.15}$$

where $\{F_e^g\}$ is the member force vector in the global coordinate system and $[T]$ is the transformation matrix, both given as

$$\{F_e^g\} = \begin{Bmatrix} H_i^g \\ V_i^g \\ M_i^g \\ H_j^g \\ V_j^g \\ M_j^g \end{Bmatrix} \text{ and } [T] = \begin{bmatrix} \cos\alpha & -\sin\alpha & 0 & 0 & 0 & 0 \\ \sin\alpha & \cos\alpha & 0 & 0 & 0 & 0 \\ 0 & 0 & 1 & 0 & 0 & 0 \\ 0 & 0 & 0 & \cos\alpha & -\sin\alpha & 0 \\ 0 & 0 & 0 & \sin\alpha & \cos\alpha & 0 \\ 0 & 0 & 0 & 0 & 0 & 1 \end{bmatrix}.$$

1.5.2 Displacement Transformation

The displacements in the global coordinate system can be related to those in the local coordinate system by following the procedure similar to the force transformation. The displacements in both coordinate systems are shown in Figure 1.13.

From Figure 1.13,

$$u_i = u_i^g \cos\alpha + v_i^g \sin\alpha \tag{1.16a}$$

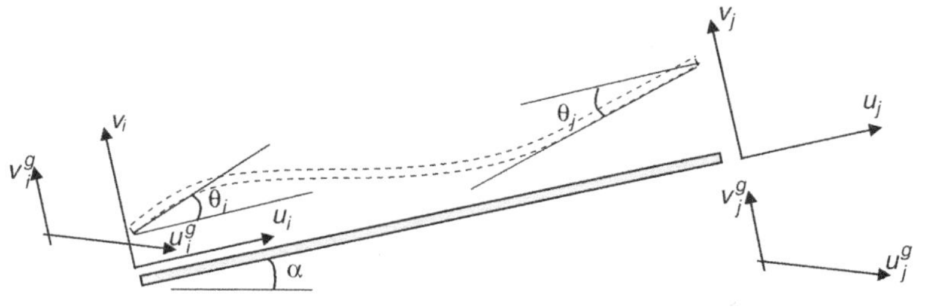

FIGURE 1.13. Displacements in the local and global coordinate systems.

$$v_i = -u_i^g \sin\alpha + v_i^g \cos\alpha \tag{1.16b}$$

$$\theta_i = \theta_i^g \tag{1.16c}$$

$$u_j = u_j^g \cos\alpha + v_j^g \sin\alpha \tag{1.16d}$$

$$v_j = -u_j^g \sin\alpha + v_j^g \cos\alpha \tag{1.16e}$$

$$\theta_j = \theta_j^g \tag{1.16f}$$

In matrix form, Equations (1.16a) to (1.16f) can be expressed as

$$\{d\} = [T]^t\{D_e^g\} \tag{1.17}$$

where $\{D_e^g\}$ is the member displacement vector in the global coordinate system corresponding to the directions in which the freedom codes are specified and is given as

$$\{D_e^g\} = \begin{Bmatrix} u_i^g \\ v_i^g \\ \theta_i^g \\ u_j^g \\ v_j^g \\ \theta_j^g \end{Bmatrix}$$

and $[T]^t$ is the transpose of $[T]$.

1.6 Member Stiffness Matrix in Global Coordinate System

From Equation (1.15),

$$\{F_e^g\} = [T]\{P\}$$

$$= [T][K_e]\{d\} \qquad \text{from Equation (1.9)}$$

$$= [T][K_e][T]^t\{D_e^g\} \quad \text{from Equation (1.17)}$$

$$= [K_e^g]\{D_e^g\} \tag{1.18}$$

where $[K_e^g] = [T][K_e][T]^t$ = member stiffness matrix in the global coordinate system.

An explicit expression for $[K_e^g]$ is

$$[K_e^g] = \begin{bmatrix} C^2\frac{EA}{L} + S^2\frac{12EI}{L^3} & SC\left(\frac{EA}{L} - \frac{12EI}{L^3}\right) & -S\frac{6EI}{L^2} & -\left(C^2\frac{EA}{L} + S^2\frac{12EI}{L^3}\right) & -SC\left(\frac{EA}{L} - \frac{12EI}{L^3}\right) & -S\frac{6EI}{L^2} \\ & S^2\frac{EA}{L} + C^2\frac{12EI}{L^3} & C\frac{6EI}{L^2} & -SC\left(\frac{EA}{L} - \frac{12EI}{L^3}\right) & -\left(S^2\frac{EA}{L} + C^2\frac{12EI}{L^3}\right) & C\frac{6EI}{L^2} \\ & & \frac{4EI}{L} & S\frac{6EI}{L^2} & -C\frac{6EI}{L^2} & \frac{2EI}{L} \\ & & & C^2\frac{EA}{L} + S^2\frac{12EI}{L^3} & SC\left(\frac{EA}{L} - \frac{12EI}{L^3}\right) & S\frac{6EI}{L^2} \\ & \text{Symmetric} & & & S^2\frac{EA}{L} + C^2\frac{12EI}{L^3} & -C\frac{6EI}{L^2} \\ & & & & & \frac{4EI}{L} \end{bmatrix} \tag{1.19}$$

where C = cos α; S = sin α.

1.7 Assembly of Structure Stiffness Matrix

Consider part of a structure with four externally applied forces, F1, F2, F4, and F5, and two applied moments, M3 and M6, acting at the two joints p and q connecting three members A, B, and C as shown in Figure 1.14. The freedom codes at joint p are {1, 2, 3} and at joint q are {4, 5, 6}. The structure stiffness matrix [K] is assembled on the basis of two conditions: compatibility and equilibrium conditions at the joints.

1.7.1 Compatibility Condition

At joint p, the global displacements are $D1$ (horizontal), $D2$ (vertical), and $D3$ (rotational). Similarly, at joint q, the global displacements are $D4$ (horizontal), $D5$ (vertical), and $D6$ (rotational). The compatibility condition is that the displacements ($D1$, $D2$, and $D3$) at end p of member A are the same as those at end p of member B. Thus, $(u_j^g)_A = (u_i^g)_B = D1$, $(v_j^g)_A = (v_i^g)_B = D2$, and $(\theta_j^g)_A = (\theta_i^g)_B = D3$. The same condition applies to displacements ($D4$, $D5$, and $D6$) at end q of both members B and C.

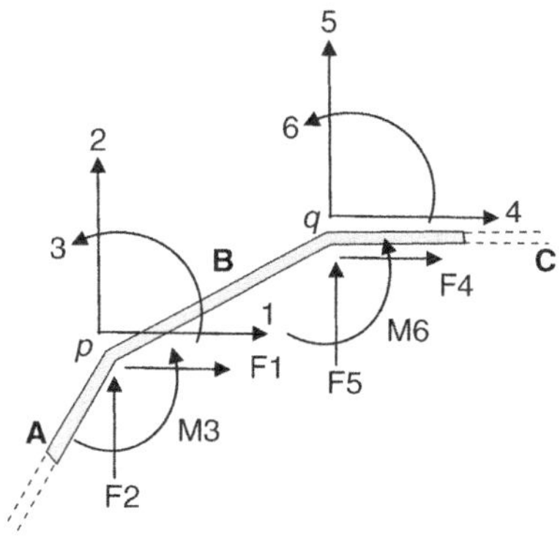

FIGURE 1.14. Assembly of structure stiffness matrix $[K]$.

The member stiffness matrix in the global coordinate system given in Equation (1.19) can be written as

$$[K_e^g] = \begin{bmatrix} k_{11} & k_{12} & k_{13} & k_{14} & k_{15} & k_{16} \\ k_{21} & k_{22} & k_{23} & k_{24} & k_{25} & k_{26} \\ k_{31} & k_{32} & k_{33} & k_{34} & k_{35} & k_{36} \\ k_{41} & k_{42} & k_{43} & k_{44} & k_{45} & k_{46} \\ k_{51} & k_{52} & k_{53} & k_{54} & k_{55} & k_{56} \\ k_{61} & k_{62} & k_{63} & k_{64} & k_{65} & k_{66} \end{bmatrix} \tag{1.20}$$

where $k_{11} = C^2 \dfrac{EA}{L} + S^2 \dfrac{12EI}{L^3}$, etc.

For member A, from Equation (1.18),

$$\left(H_j^g\right)_A = \ldots\ldots + \ldots\ldots + \ldots\ldots + (k_{44})_A D_1 + (k_{45})_A D_2 + (k_{46})_A D_3 \tag{1.21a}$$

$$\left(V_j^g\right)_A = \ldots\ldots + \ldots\ldots + \ldots\ldots + (k_{54})_A D_1 + (k_{55})_A D_2 + (k_{56})_A D_3 \tag{1.21b}$$

$$\left(M_j^g\right)_A = \ldots\ldots + \ldots\ldots + \ldots\ldots + (k_{64})_A D_1 + (k_{65})_A D_2 + (k_{66})_A D_3 \tag{1.21c}$$

Similarly, for member B,

$$(H_i^g)_B = (k_{11})_B D_1 + (k_{12})_B D_2 + (k_{13})_B D_3 + (k_{14})_B D_4 + (k_{15})_B D_5 + (k_{16})_B D_6 \tag{1.21d}$$

$$(V_i^g)_B = (k_{21})_B D_1 + (k_{22})_B D_2 + (k_{23})_B D_3 + (k_{24})_B D_4 + (k_{25})_B D_5 + (k_{26})_B D_6 \tag{1.21e}$$

$$(M_i^g)_B = (k_{31})_B D_1 + (k_{32})_B D_2 + (k_{33})_B D_3 + (k_{34})_B D_4 + (k_{35})_B D_5 + (k_{36})_B D_6 \tag{1.21f}$$

$$\left(H_j^g\right)_B = (k_{41})_B D_1 + (k_{42})_B D_2 + (k_{43})_B D_3 + (k_{44})_B D_4 + (k_{45})_B D_5 + (k_{46})_B D_6 \tag{1.21g}$$

$$\left(V_j^g\right)_B = (k_{51})_B D_1 + (k_{52})_B D_2 + (k_{53})_B D_3 + (k_{54})_B D_4 + (k_{55})_B D_5 + (k_{56})_B D_6 \tag{1.21h}$$

$$\left(M_j^g\right)_B = (k_{61})_B D_1 + (k_{62})_B D_2 + (k_{63})_B D_3 + (k_{64})_B D_4 + (k_{65})_B D_5 + (k_{66})_B D_6 \tag{1.21i}$$

Similarly, for member C,

$$(H_i^g)_C = (k_{11})_C D_1 + (k_{12})_C D_2 + (k_{13})_C D_3 + \ldots\ldots + \ldots\ldots + \ldots\ldots \tag{1.21j}$$

$$(V_i^g)_C = (k_{21})_C D_1 + (k_{22})_C D_2 + (k_{23})_C D_3 + \ldots\ldots + \ldots\ldots + \ldots\ldots \tag{1.21k}$$

$$(M_i^g)_C = (k_{31})_C D_1 + (k_{32})_C D_2 + (k_{33})_C D_3 + \ldots\ldots + \ldots\ldots + \ldots\ldots \tag{1.21l}$$

1.7.2 Equilibrium Condition

Any of the externally applied forces or moments applied in a certain direction at a joint of a structure is equal to the sum of the member forces acting in the same direction for members connected at that joint in the global coordinate system. Therefore, at joint p,

$$F1 = \left(H_j^g\right)_A + (H_i^g)_B \tag{1.22a}$$

$$F2 = \left(V_j^g\right)_A + (V_i^g)_B \tag{1.22b}$$

$$M3 = \left(M_j^g\right)_A + (M_i^g)_B \tag{1.22c}$$

Also, at joint q,

$$F4 = \left(H_j^g\right)_B + (H_i^g)_C \tag{1.22d}$$

$$F5 = \left(V_j^g\right)_B + (V_i^g)_C \tag{1.22e}$$

$$M6 = \left(M_j^g\right)_B + (M_i^g)_C \tag{1.22f}$$

By writing Equations (1.22a) to (1.22f) in matrix form using Equations (1.21a) to (1.21l) and applying this operation to the whole structure, the following equilibrium equation of the whole structure is obtained:

$$
\begin{Bmatrix} \bullet \\ F1 \\ F2 \\ M3 \\ F4 \\ F5 \\ M6 \\ \bullet \end{Bmatrix}
=
\begin{bmatrix}
\bullet & \bullet & \bullet & \bullet & \bullet & \bullet & \bullet & \bullet \\
\bullet & (k_{44})_A + (k_{11})_B & (k_{45})_A + (k_{12})_B & (k_{46})_A + (k_{13})_B & (k_{14})_B & (k_{15})_B & (k_{16})_B & \bullet \\
\bullet & (k_{54})_A + (k_{21})_B & (k_{55})_A + (k_{22})_B & (k_{56})_A + (k_{23})_B & (k_{24})_B & (k_{25})_B & (k_{26})_B & \bullet \\
\bullet & (k_{64})_A + (k_{31})_B & (k_{65})_A + (k_{32})_B & (k_{66})_A + (k_{33})_B & (k_{34})_B & (k_{35})_B & (k_{36})_B & \bullet \\
\bullet & (k_{41})_B & (k_{42})_B & (k_{43})_B & (k_{44})_B + (k_{11})_C & (k_{45})_B + (k_{12})_C & (k_{46})_B + (k_{13})_C & \bullet \\
\bullet & (k_{51})_B & (k_{52})_B & (k_{53})_B & (k_{54})_B + (k_{21})_C & (k_{55})_B + (k_{22})_C & (k_{56})_B + (k_{23})_C & \bullet \\
\bullet & (k_{61})_B & (k_{62})_B & (k_{63})_B & (k_{64})_B + (k_{31})_C & (k_{65})_B + (k_{32})_C & (k_{66})_B + (k_{33})_C & \bullet \\
\bullet & \bullet & \bullet & \bullet & \bullet & \bullet & \bullet & \bullet
\end{bmatrix}
\begin{Bmatrix} \bullet \\ \bullet \\ D1 \\ D2 \\ D3 \\ D4 \\ D5 \\ D4 \\ \bullet \\ \bullet \end{Bmatrix}
\tag{1.23}
$$

where the "•" stands for matrix coefficients contributed from the other parts of the structure. In simple form, Equation (1.23) can be written as

$$\{F\} = [K]\{D\}$$

which is identical to Equation (1.7). Equation (1.23) shows how the structure equilibrium equation is set up in terms of the load vector $\{F\}$, structure stiffness matrix $[K]$, and the displacement vector $\{D\}$.

Close examination of Equation (1.23) reveals that the stiffness coefficients of the three members A, B, and C are assembled into $[K]$ in a way according to the freedom codes assigned to the members. Take member A as an example. By writing the freedom codes in the order of ends i and j around the member stiffness matrix in the global coordinate system shown in Figure 1.15, the coefficient, for example, k_{54}, is assembled into the position [2, 1] of $[K]$. Similarly, the coefficient k_{45} is assembled into the position [1, 2] of $[K]$. The coefficients in all member stiffness matrices in the global coordinate system can be assembled into $[K]$ in this way. Since the resulting matrix is symmetric, only half of the coefficients need to be assembled.

A schematic diagram showing the assembly procedure for the stiffness coefficients of the three members A, B, and C into $[K]$ is shown in Figure 1.16. Note that since $[K_e^g]$ is symmetric, $[K]$ is also symmetric. Any coefficients in a row or column corresponding to zero freedom code will be ignored.

1.8 Load Vector

The load vector $\{F\}$ of a structure is formed by assembling the individual forces into the load vector in positions corresponding to the directions of the freedom codes. For the example in Figure 1.14, the load factor is given as that shown in Figure 1.17.

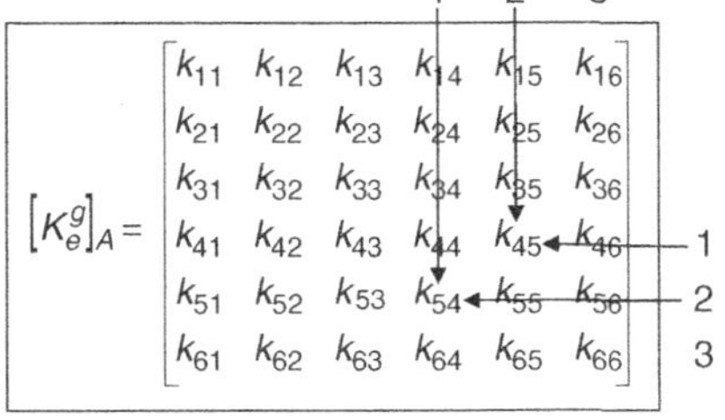

FIGURE 1.15. Assembly of stiffness coefficients into the structure stiffness matrix.

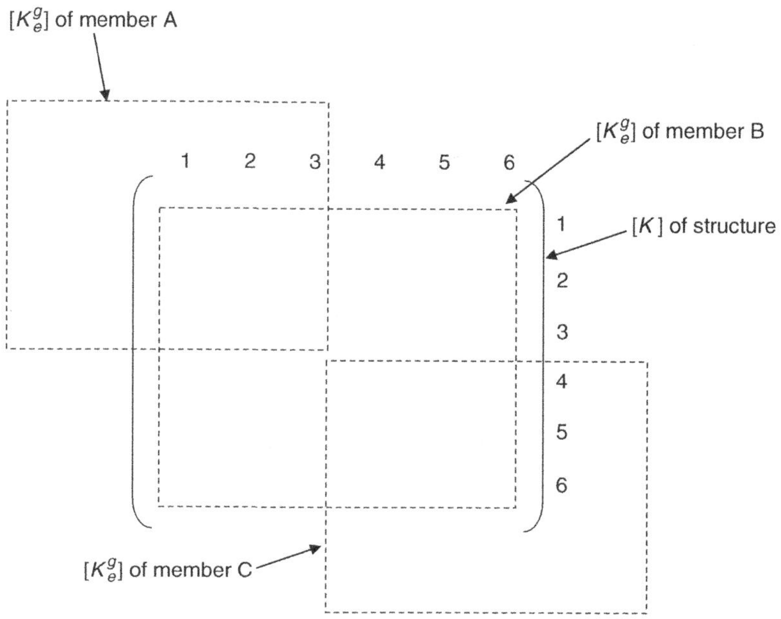

FIGURE 1.16. Assembly of structure stiffness matrix.

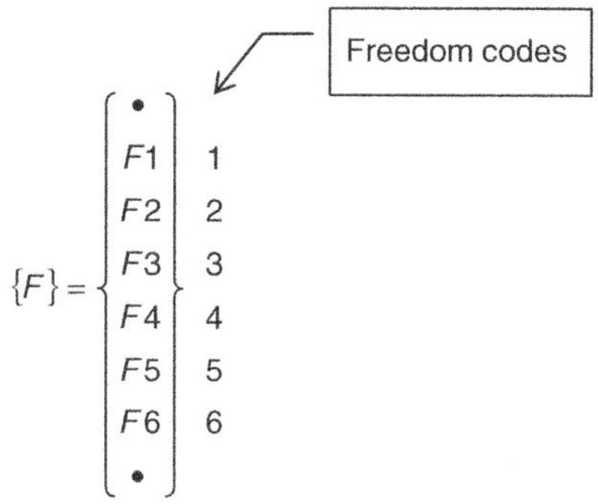

FIGURE 1.17. Assembly of load vector.

1.9 Methods of Solution

The displacements of the structure can be found by solving Equation (1.23). Because of the huge size of the matrix equation usually encountered in practice, Equation (1.23) is solved routinely by numerical methods such as the Gaussian elimination method and the iterative Gauss–Seidel method. It should be noted that in using these numerical methods, the procedure is analogous to inverting the structure stiffness matrix, which is subsequently multiplied by the load vector as in Equation (1.8):

$$\{D\} = [K]^{-1}\{F\} \tag{1.8}$$

The numerical procedure fails only if an inverted [K] cannot be found. This situation occurs when the determinant of [K] is zero, implying an unstable structure. Unstable structures with a degree of statically indeterminacy, f_r, greater than zero (see Section 1.2) will have a zero determinant of [K]. In numerical manipulation by computers, an exact zero is sometimes difficult to obtain. In such cases, a good indication of an unstable structure is to examine the displacement vector {D}, which would include some exceptionally large values.

1.10 Calculation of Member Forces

Member forces are calculated according to Equation (1.9). Hence,

$$\begin{aligned}\{P\} &= [K_e]\{d\} \\ &= [K_e]\,[T]^t\{D_e^g\}\end{aligned} \tag{1.24}$$

where $\{D_e^g\}$ is extracted from {D} for each member according to its freedom codes and

$$[K_e][T]^t = \begin{bmatrix} C\frac{EA}{L} & S\frac{EA}{L} & 0 & -C\frac{EA}{L} & -S\frac{EA}{L} & 0 \\ -S\frac{12EI}{L^3} & C\frac{12EI}{L^3} & \frac{6EI}{L^2} & S\frac{12EI}{L^3} & -C\frac{12EI}{L^3} & \frac{6EI}{L^2} \\ -S\frac{6EI}{L^2} & C\frac{6EI}{L^2} & \frac{4EI}{L} & S\frac{6EI}{L^2} & -C\frac{6EI}{L^2} & \frac{2EI}{L} \\ -C\frac{EA}{L} & -S\frac{EA}{L} & 0 & C\frac{EA}{L} & S\frac{EA}{L} & 0 \\ S\frac{12EI}{L^3} & -C\frac{12EI}{L^3} & -\frac{6EI}{L^2} & -S\frac{12EI}{L^3} & C\frac{12EI}{L^3} & -\frac{6EI}{L^2} \\ -S\frac{6EI}{L^2} & C\frac{6EI}{L^2} & \frac{2EI}{L} & S\frac{6EI}{L^2} & -C\frac{6EI}{L^2} & \frac{4EI}{L} \end{bmatrix}$$

For the example in Figure 1.14,

$$\{P\} = \begin{Bmatrix} N_i \\ Q_i \\ M_i \\ N_j \\ Q_j \\ M_j \end{Bmatrix} = \begin{bmatrix} C\frac{EA}{L} & S\frac{EA}{L} & 0 & -C\frac{EA}{L} & -S\frac{EA}{L} & 0 \\ -S\frac{12EI}{L^3} & C\frac{12EI}{L^3} & \frac{6EI}{L^2} & S\frac{12EI}{L^3} & -C\frac{12EI}{L^3} & \frac{6EI}{L^2} \\ -S\frac{6EI}{L^2} & C\frac{6EI}{L^2} & \frac{4EI}{L} & S\frac{6EI}{L^2} & -C\frac{6EI}{L^2} & \frac{2EI}{L} \\ -C\frac{EA}{L} & -S\frac{EA}{L} & 0 & C\frac{EA}{L} & S\frac{EA}{L} & 0 \\ S\frac{12EI}{L^3} & -C\frac{12EI}{L^3} & -\frac{6EI}{L^2} & -S\frac{12EI}{L^3} & C\frac{12EI}{L^3} & -\frac{6EI}{L^2} \\ -S\frac{6EI}{L^2} & C\frac{6EI}{L^2} & \frac{2EI}{L} & S\frac{6EI}{L^2} & -C\frac{6EI}{L^2} & \frac{4EI}{L} \end{bmatrix} \begin{Bmatrix} D_1 \\ D_2 \\ D_3 \\ D_4 \\ D_5 \\ D_6 \end{Bmatrix}$$

In summary, the procedure for using the stiffness method to calculate the displacements of the structure and the member forces is as follows.

1. Assign freedom codes to each joint indicating the displacement freedom at the ends of the members connected at that joint. Assign a freedom code of "zero" to any restrained displacement.
2. Assign an arrow to each member so that ends i and j are defined. Also, the angle of orientation α for the member is defined in Figure 1.18 as:

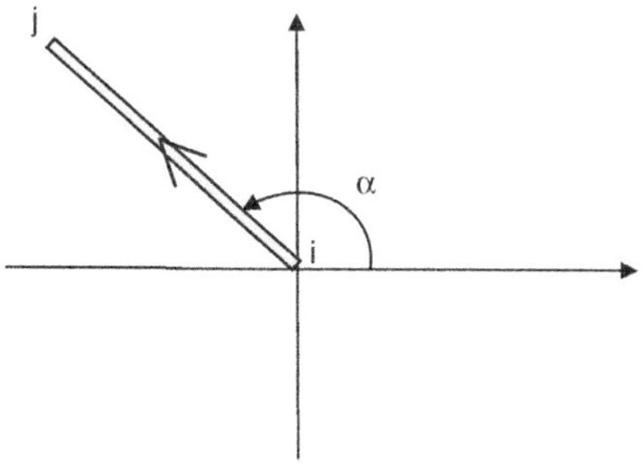

FIGURE 1.18. Definition of angle of orientation for member.

3. Assemble the structure stiffness matrix $[K]$ from each of the member stiffness matrices.
4. Form the load vector $\{F\}$ of the structure.

5. Calculate the displacement vector $\{D\}$ by solving for $\{D\} = [K]^{-1}\{F\}$.
6. Extract the local displacement vector $\{D_e^g\}$ from $\{D\}$ and calculate the member force vector $\{P\}$ using $\{P\} = [K_e][T]^t\{D_e^g\}$.

1.10.1 Sign Convention for Member Force Diagrams

Positive member forces and displacements obtained from the stiffness method of analysis are shown in Figure 1.19. To plot the forces in conventional axial force, shear force, and bending moment diagrams, it is necessary to translate them into a system commonly adopted for plotting.

The sign convention for such a system is given as follows.

Axial Force

For a member under compression, the axial force at end i is positive (from analysis) and at end j is negative (from analysis), as shown in Figure 1.20.

Shear Force

A shear force plotted positive in diagram is acting upward (positive from analysis) at end i and downward (negative from analysis) at end j as shown in Figure 1.21. Positive shear force is usually plotted in the space above the member.

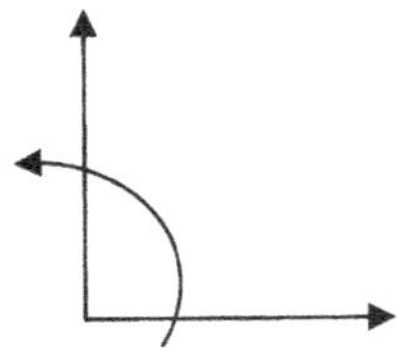

FIGURE 1.19. Direction of positive forces and displacements using stiffness method.

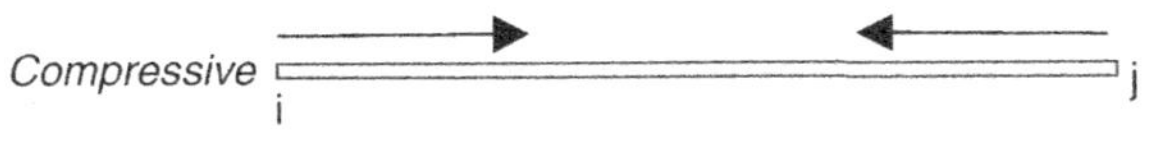

FIGURE 1.20. Member under compression.

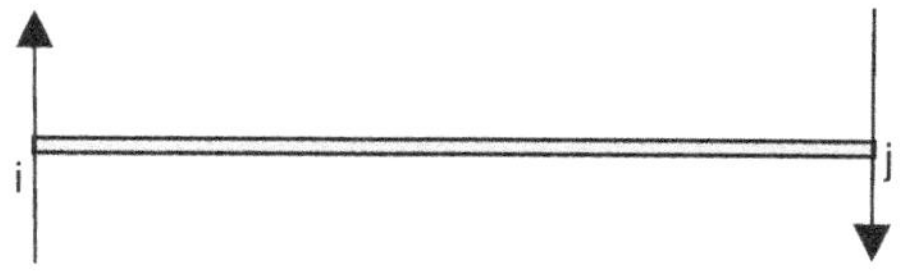

FIGURE 1.21. Positive shear forces.

Bending Moment

A member under sagging moment is positive in diagram (clockwise and negative from analysis) at end *i* and positive (anticlockwise and positive from analysis) at end *j* as shown in Figure 1.22. Positive bending moment is usually plotted in the space beneath the member. In doing so, a bending moment is plotted on the tension face of the member.

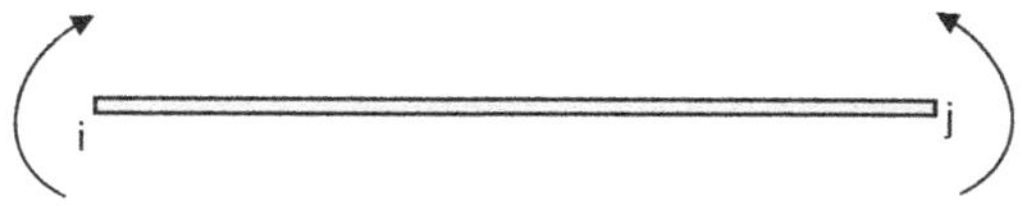

FIGURE 1.22. Sagging moment of a member.

Example 1.3 Determine the member forces and plot the shear force and bending moment diagrams for the structure shown in Figure 1.23a. The structure with a pin at D is subject to a vertical force of 100 kN being applied at C. For all members, $E = 2 \times 10^8$ kN/m^2, $A = 0.2$ m^2, and $I = 0.001$ m^4.

Solution. The freedom codes for the whole structure are shown in Figure 1.23b. There are four members separated by joints B, C, and D with the member numbers shown. The arrows are assigned to

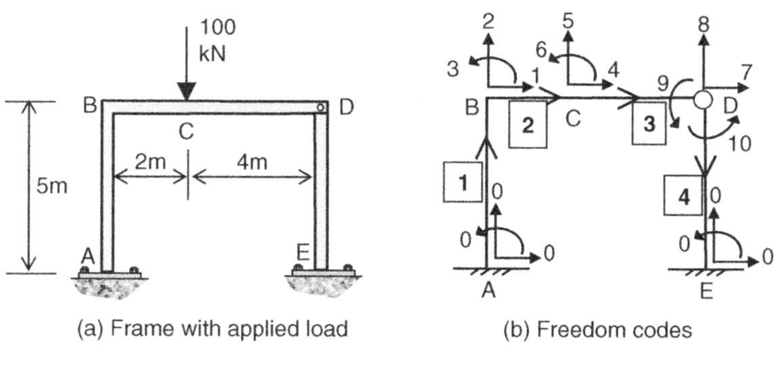

FIGURE 1.23. Example 1.3.

indicate end i (tail of arrow) and end j (head of arrow). Thus, the orientations of the members are

Member 1: $\alpha = 90^\circ$
Member 2: $\alpha = 0^\circ$
Member 3: $\alpha = 0^\circ$
Member 4: $\alpha = 270^\circ$ or -90°

The $[K_e^g]$ for the members with the assigned freedom codes for the coefficients is

$$[K_e^g]_1 = \begin{array}{cccccc|c}
0 & 0 & 0 & 1 & 2 & 3 & \\
\hline
1.92 \times 10^4 & 0 & -4.8 \times 10^4 & -1.92 \times 10^4 & 0 & -4.8 \times 10^4 & 0 \\
 & 8 \times 10^6 & 0 & 0 & -8 \times 10^6 & 0 & 0 \\
 & & 1.6 \times 10^5 & 4.8 \times 10^4 & 0 & 8 \times 10^4 & 0 \\
 & & & 1.92 \times 10^4 & 0 & 4.8 \times 10^4 & 1 \\
 & \text{Symmetric} & & & 8 \times 10^6 & 0 & 2 \\
 & & & & & 1.6 \times 10^5 & 3
\end{array}$$

$$[K_e^g]_2 = \begin{array}{cccccc|c}
1 & 2 & 3 & 4 & 5 & 6 & \\
\hline
2 \times 10^7 & 0 & 0 & -2 \times 10^7 & 0 & 0 & 1 \\
 & 3 \times 10^5 & 3 \times 10^5 & 0 & -3 \times 10^5 & 3 \times 10^5 & 2 \\
 & & 4 \times 10^5 & 0 & -3 \times 10^5 & 2 \times 10^5 & 3 \\
 & & & 2 \times 10^7 & 0 & 0 & 4 \\
 & \text{Symmetric} & & & 3 \times 10^5 & -3 \times 10^5 & 5 \\
 & & & & & 4 \times 10^5 & 6
\end{array}$$

$$[K_e^g]_3 = \begin{array}{cccccc|c}
4 & 5 & 6 & 7 & 8 & 9 & \\
\hline
1 \times 10^7 & 0 & 0 & -1 \times 10^7 & 0 & 0 & 4 \\
 & 3.75 \times 10^4 & 7.5 \times 10^4 & 0 & -3.75 \times 10^4 & 7.5 \times 10^4 & 5 \\
 & & 2 \times 10^5 & 0 & -7.5 \times 10^4 & 1 \times 10^5 & 6 \\
 & & & 1 \times 10^7 & 0 & 0 & 7 \\
 & \text{Symmetric} & & & 3.75 \times 10^4 & -7.5 \times 10^4 & 8 \\
 & & & & & 2 \times 10^5 & 9
\end{array}$$

$$[K_e^g]_4 = \begin{array}{cccccc|c}
7 & 8 & 10 & 0 & 0 & 0 & \\
\hline
1.92 \times 10^4 & 0 & -4.8 \times 10^4 & -1.92 \times 10^4 & 0 & 4.8 \times 10^4 & 7 \\
 & 8 \times 10^6 & 0 & 0 & -8 \times 10^6 & 0 & 8 \\
 & & 1.6 \times 10^5 & -4.8 \times 10^4 & 0 & 8 \times 10^4 & 10 \\
 & & & 1.92 \times 10^4 & 0 & -4.8 \times 10^4 & 0 \\
 & \text{Symmetric} & & & 8 \times 10^6 & 0 & 0 \\
 & & & & & 1.6 \times 10^5 & 0
\end{array}$$

By assembling from $[K_e^g]$ of all members, the structure stiffness matrix is obtained:

$$[K] = \begin{bmatrix}
2.0019 \times 10^7 & 0 & 4.8 \times 10^4 & -2 \times 10^7 & 0 & 0 & 0 & 0 & 0 & 0 \\
 & 8.3 \times 10^6 & 3 \times 10^5 & 0 & -3 \times 10^5 & 3 \times 10^5 & 0 & 0 & 0 & 0 \\
 & & 5.6 \times 10^5 & 0 & -3 \times 10^5 & 2 \times 10^5 & 0 & 0 & 0 & 0 \\
 & & & 3 \times 10^7 & 0 & 0 & -1 \times 10^7 & 0 & 0 & 0 \\
 & & & & 3.375 \times 10^5 & -2.25 \times 10^5 & 0 & -3.75 \times 10^4 & 7.5 \times 10^4 & 0 \\
 & \textit{Symmetric} & & & & 6 \times 10^5 & 0 & -7.5 \times 10^4 & 1 \times 10^5 & 0 \\
 & & & & & & 1.0019 \times 10^7 & 0 & 0 & 4.8 \times 10^4 \\
 & & & & & & & 8.0375 \times 10^6 & -7.5 \times 10^4 & 0 \\
 & & & & & & & & 2 \times 10^5 & 0 \\
 & & & & & & & & & 1.6 \times 10^5
\end{bmatrix}$$

The load vector is given by

$$\{F\} = \begin{Bmatrix} 0 \\ 0 \\ 0 \\ 0 \\ -100 \\ 0 \\ 0 \\ 0 \\ 0 \\ 0 \end{Bmatrix} \text{ and } \{D\} = [K]^{-1}\{F\} = \begin{Bmatrix} 1.354 \times 10^{-3} \\ -9.236 \times 10^{-6} \\ -6.770 \times 10^{-4} \\ 1.354 \times 10^{-3} \\ -1.304 \times 10^{-3} \\ -3.713 \times 10^{-4} \\ 1.353 \times 10^{-3} \\ -3.264 \times 10^{-6} \\ 6.733 \times 10^{-4} \\ -4.059 \times 10^{-4} \end{Bmatrix} \begin{matrix} \text{m} \\ \text{m} \\ \text{radian} \\ \text{m} \\ \text{m} \\ \text{radian} \\ \text{m} \\ \text{m} \\ \text{radian} \\ \text{radian} \end{matrix}$$

The member forces can be calculated using $\{P\} = [K_e][T]^t\{D_e^g\}$.

For member 1, where C = cos 90° = 0 and S = sin 90° = 1,

$$\{P\}_1 = \begin{Bmatrix} N_i \\ Q_i \\ M_i \\ N_j \\ Q_j \\ M_j \end{Bmatrix} = \begin{bmatrix} C\frac{EA}{L} & S\frac{EA}{L} & 0 & -C\frac{EA}{L} & -S\frac{EA}{L} & 0 \\ -S\frac{12EI}{L^3} & C\frac{12EI}{L^3} & \frac{6EI}{L^2} & S\frac{12EI}{L^3} & -C\frac{12EI}{L^3} & \frac{6EI}{L^2} \\ -S\frac{6EI}{L^2} & C\frac{6EI}{L^2} & \frac{4EI}{L} & S\frac{6EI}{L^2} & -C\frac{6EI}{L^2} & \frac{2EI}{L} \\ -C\frac{EA}{L} & -S\frac{EA}{L} & 0 & C\frac{EA}{L} & S\frac{EA}{L} & 0 \\ S\frac{12EI}{L^3} & -C\frac{12EI}{L^3} & -\frac{6EI}{L^2} & -S\frac{12EI}{L^3} & C\frac{12EI}{L^3} & -\frac{6EI}{L^2} \\ -S\frac{6EI}{L^2} & C\frac{6EI}{L^2} & \frac{2EI}{L} & S\frac{6EI}{L^2} & -C\frac{6EI}{L^2} & \frac{4EI}{L} \end{bmatrix} \begin{Bmatrix} 0 \\ 0 \\ 0 \\ 1.354 \times 10^{-3} \\ -9.236 \times 10^{-6} \\ -6.770 \times 10^{-4} \end{Bmatrix}$$

$$= \begin{Bmatrix} 73.9 \\ -6.5 \\ 10.8 \\ -73.9 \\ 6.5 \\ -43.3 \end{Bmatrix} \begin{matrix} \text{kN} \\ \text{kN} \\ \text{kNm} \\ \text{kN} \\ \text{kN} \\ \text{kNm} \end{matrix}$$

Similarly,

$$\{P\}_2 = \begin{Bmatrix} 6.5 \\ 73.9 \\ 43.3 \\ -6.5 \\ -73.9 \\ 104.5 \end{Bmatrix} \begin{matrix} \text{kN} \\ \text{kN} \\ \text{kNm} \\ \text{kN} \\ \text{kN} \\ \text{kNm} \end{matrix}, \{P\}_3 = \begin{Bmatrix} 6.5 \\ -26.1 \\ -104.5 \\ -6.5 \\ 26.1 \\ 0 \end{Bmatrix} \begin{matrix} \text{kN} \\ \text{kN} \\ \text{kNm} \\ \text{kN} \\ \text{kN} \\ \text{kNm} \end{matrix}, \text{ and } \{P\}_4 = \begin{Bmatrix} 26.1 \\ 6.5 \\ 0 \\ -26.1 \\ -6.5 \\ 32.5 \end{Bmatrix} \begin{matrix} \text{kN} \\ \text{kN} \\ \text{kNm} \\ \text{kN} \\ \text{kN} \\ \text{kNm} \end{matrix}$$

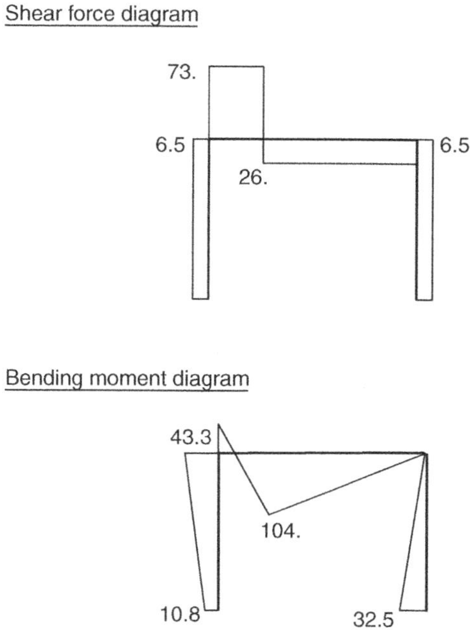

FIGURE 1.24. Shear force and bending moment diagrams.

The shear force and bending moment diagrams are shown in Figure 1.24.

1.11 Treatment of Internal Loads

So far, the discussion has concerned externally applied loads acting only at joints of the structure. However, in many instances, externally applied loads are also applied at locations other than the joints, such as on part or whole of a member. Loads being applied in this manner are termed internal loads. Internal loads may include distributed loads, point loads, and loads due to temperature effects. In such cases, the loads are calculated by treating the member as fixed-end, and fixed-end forces, including axial forces, shear forces, and bending moments, are calculated at its ends. The fictitiously fixed ends of the member are then removed and the effects of the fixed-end forces, now being treated as applied loads at the joints, are assessed using the stiffness method of analysis.

In Figure 1.25, fixed-end forces due to the point load and the uniformly distributed load are collected in a fixed-end force vector $\{P_F\}$ for the member as

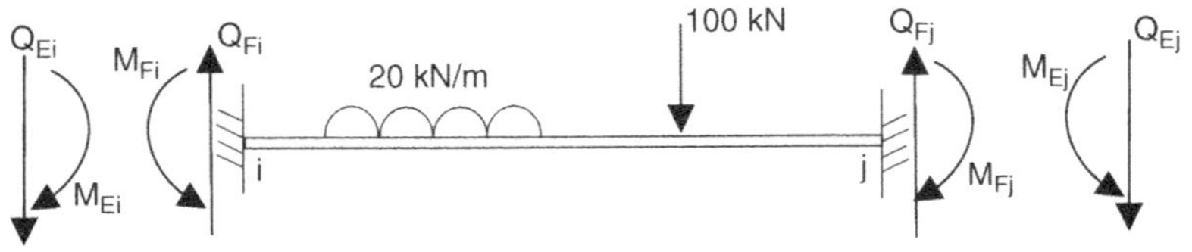

FIGURE 1.25. Fixed-end forces.

$$\{P_F\} = \begin{Bmatrix} 0 \\ Q_{Fi} \\ M_{Fi} \\ 0 \\ Q_{Fj} \\ M_{Fj} \end{Bmatrix} \tag{1.25}$$

The signs of the forces in $\{P_F\}$ should follow those shown in Figure 1.19. In equilibrium, fixed-end forces generate a set of equivalent forces, equal in magnitude but opposite in sense and shown as Q_{Ei}, M_{Ei}, Q_{Ej}, M_{Ej}, being applied at the joints pertaining to both ends i and j of the member. The equivalent force vector is expressed as

$$\{P_E\} = \begin{Bmatrix} 0 \\ -Q_{Fi} \\ -M_{Fi} \\ 0 \\ -Q_{Fj} \\ -M_{Fj} \end{Bmatrix} \tag{1.26}$$

If necessary, $\{P_E\}$ is transformed into the global coordinate system in a similar way given in Equation (1.15) to form

$$\{P_E^g\} = [T]\{P_E\} \tag{1.27}$$

which is added to the load vector $\{F\}$ of the structure in accordance with the freedom codes at the joints. Final member forces are calculated as the sum of the forces obtained from the global structural analysis and fixed-end forces $\{P_F\}$. That is,

$$\{P\} = [K_e]\{d\} + \{P_F\} \tag{1.28}$$

Fixed-end forces for two common loading cases are shown in Table 1.1.

Example 1.4 Determine the forces in the members and plot the bending moment and shear force diagrams for the frame shown in Figure 1.26a. The structure is fixed at A and pinned on a roller support

TABLE 1.1
Fixed-end forces

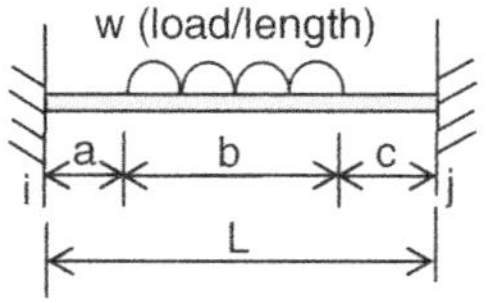	Shear force at end i, Q_{Fi}	$Q_{Fi} = \frac{wb(\frac{b}{2}+c)}{L} + \frac{M_{Fi}+M_{Fj}}{L}$
	Shear force at end j, Q_{Fj}	$Q_{Fj} = wb - Q_{Fi}$
	Bending moment at end i, M_{Fi}	$M_{Fi} = \frac{w}{12L^2}\left[(L-a)^3(L+3a) - c^3(4L-3c)\right]$
	Bending moment at end j, M_{Fj}	$M_{Fj} = -\frac{w}{12L^2}\left[(L-c)^3(L+3c) - a^3(4L-3a)\right]$
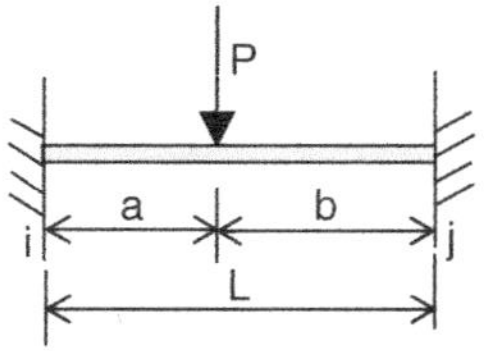	Shear force at end i, Q_{Fi}	$Q_{Fi} = P\left(\frac{b}{L}\right)^2\left(1+\frac{2a}{L}\right)$
	Shear force at end j, Q_{Fj}	$Q_{Fj} = P - Q_{Fi}$
	Bending moment at end i, M_{Fi}	$M_{Fj} = \frac{Pab^2}{L^2}$
	Bending moment at end j, M_{Fj}	$M_{Fj} = \frac{Pa^2b}{L^2}$

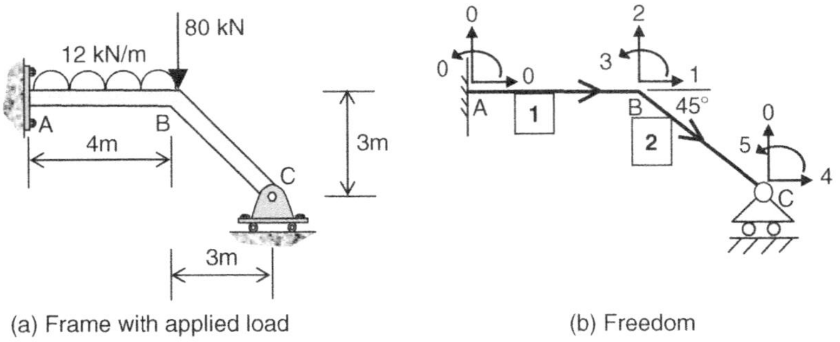

FIGURE 1.26. Example 1.4.

at C. For both members AB and BC, $E = 2 \times 10^8$ kN/m^2, $A = 0.2$ m^2, $I = 0.001$ m^4.

Solution. The structure has 5 degrees of freedom with a degree of statical indeterminacy of 2. Freedom codes corresponding to the 5 degrees of freedom are shown in Figure 1.26b. The fixed-end force vector for member 1 is

$$\{P_F\} = \begin{Bmatrix} 0 \\ 24 \\ 16 \\ 0 \\ 24 \\ -16 \end{Bmatrix}$$

The equivalent force vector is

$$\{P_E\} = \{P_E^g\} = \begin{Bmatrix} 0 \\ -24 \\ -16 \\ 0 \\ -24 \\ 16 \end{Bmatrix} \begin{matrix} 0 \\ 0 \\ 0 \\ 1 \\ 2 \\ 3 \end{matrix}$$

which is added to the externally applied force to form

$$\{F\} = \begin{Bmatrix} 0 \\ -104 \\ 16 \\ 0 \\ 0 \end{Bmatrix} \begin{matrix} 1 \\ 2 \\ 3 \\ 4 \\ 5 \end{matrix}$$

The orientations of the members are member 1: $\alpha = 0°$, member 2: $\alpha = -45°$. The stiffness matrices of the members in the global coordinate system are

$$
[K_e^g]_1 = \begin{array}{c} \\ \left[\begin{array}{cccccc}
0 & 0 & 0 & 1 & 2 & 3 \\
1\times10^7 & 0 & 0 & -1\times10^7 & 0 & 0 \\
 & 3.75\times10^4 & 7.5\times10^4 & 0 & -3.75\times10^4 & 7.5\times10^4 \\
 & & 2\times10^5 & 0 & -7.5\times10^4 & 1\times10^5 \\
 & & & 1\times10^7 & 0 & 0 \\
 & \textit{Symmetric} & & & 3.75\times10^4 & -7.5\times10^4 \\
 & & & & & 2\times10^5
\end{array}\right] \end{array}
\begin{array}{c} \\ 0 \\ 0 \\ 0 \\ 1 \\ 2 \\ 3 \end{array}
$$

$$
[K_e^g]_2 = \left[\begin{array}{cccccc}
1 & 2 & 3 & 4 & 0 & 5 \\
0.473\times10^7 & -0.4698\times10^6 & 4.714\times10^4 & -0.473\times10^7 & 4.698\times10^6 & 4.714\times10^4 \\
 & 0.473\times10^7 & -4.714\times10^4 & 4.698\times10^6 & -0.473\times10^7 & -4.714\times10^4 \\
 & & 1.886\times10^5 & -4.714\times10^4 & 4.714\times10^4 & 9.428\times10^4 \\
 & & & 0.473\times10^7 & 4.698\times10^6 & -4.714\times10^4 \\
 & \textit{Symmetric} & & & 0.473\times10^7 & 4.714\times10^4 \\
 & & & & & 1.886\times10^5
\end{array}\right]
\begin{array}{c} \\ 1 \\ 2 \\ 3 \\ 4 \\ 0 \\ 5 \end{array}
$$

Hence, the structure stiffness matrix is assembled as

$$
[K] = \begin{bmatrix}
1.473\times10^7 & -4.698\times10^6 & 4.714\times10^4 & -4.730\times10^6 & 4.714\times10^4 \\
 & 4.767\times10^6 & -2.786\times10^4 & 4.698\times10^6 & 4.714\times10^4 \\
 & & 3.886\times10^5 & -4.714\times10^4 & 9.428\times10^4 \\
 & \textit{Symmetric} & & 4.730\times10^6 & -4.714\times10^4 \\
 & & & & 1.886\times10^5
\end{bmatrix}
$$

By solving the structure equilibrium equation, the displacement vector is determined as

$$
\{D\} = \begin{Bmatrix} 0 \\ -2.017\times10^{-3} \\ -1.180\times10^{-4} \\ 2.013\times10^{-3} \\ 1.066\times10^{-3} \end{Bmatrix}
\begin{array}{l} \text{m} \\ \text{m} \\ \text{radian} \\ \text{m} \\ \text{radian} \end{array}
$$

$$
\{P\}_1 = [K_e]\{d\} + \{P_F\} = \begin{Bmatrix} 0 \\ 66.8 \\ 139.5 \\ 0 \\ -66.8 \\ 127.7 \end{Bmatrix} + \begin{Bmatrix} 0 \\ 24 \\ 16 \\ 0 \\ 24 \\ -16 \end{Bmatrix} = \begin{Bmatrix} 0 \\ 90.8 \\ 155.5 \\ 0 \\ -42.8 \\ 111.7 \end{Bmatrix}
\begin{array}{l} \text{kN} \\ \text{kN} \\ \text{kNm} \\ \text{kN} \\ \text{kN} \\ \text{kNm} \end{array}
$$

$$
\{P\}_2 = [K_e]\{d\} = \begin{Bmatrix} 26.3 \\ -26.3 \\ -111.7 \\ -26.3 \\ 26.3 \\ 0 \end{Bmatrix}
\begin{array}{l} \text{kN} \\ \text{kN} \\ \text{kNm} \\ \text{kN} \\ \text{kN} \\ \text{kNm} \end{array}
$$

The shear force and bending moment diagrams of the structure are shown in Figure 1.27.

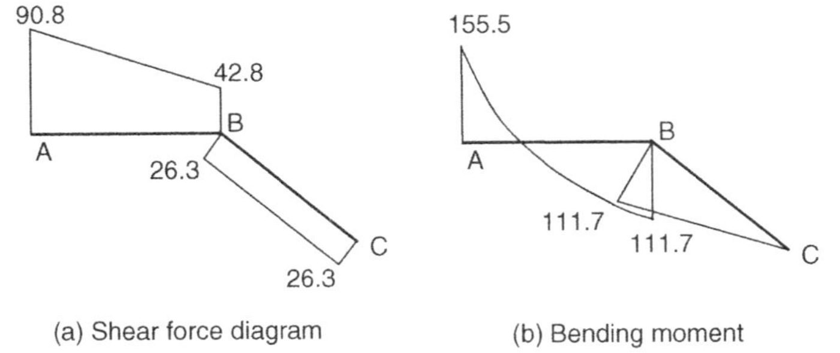

FIGURE 1.27. Results of Example 1.4.

1.12 Treatment of Pins

Example 1.3 demonstrates the analysis of a frame with a pin at joint D. The way to treat the pin using the stiffness method for structural analysis is to allow the members attached to the pinned joint to rotate independently, thus leading to the creation of different freedom codes for rotations of individual members. When carrying out elastoplastic analysis (Chapter 4) for structures using the stiffness method, the plastic hinges, behaving in a way similar to a pin, are formed in stages as the loads increase. In assigning different freedom codes to represent the creation of plastic hinges in an elastoplastic analysis, the number of degrees of freedom increases by one every time a plastic hinge is formed. For a structure with a high degree of statical indeterminacy, the increase in the number of freedom codes from the beginning of the elastoplastic analysis to its collapse due to instability induced by the formation of plastic hinges may be large. Elastoplastic analysis using this method for simulating pin behavior, hereafter called the extra freedom method, therefore requires increasing both the number of equilibrium equations to be solved and the size of the structure stiffness matrix $[K]$, thus increasing the storage requirements for the computer and decreasing the efficiency of the solution procedure. In order to maintain the size of $[K]$ and maximize computational efficiency in an elastoplastic analysis, the behavior of a pin at the ends of the member can be simulated implicitly by modifying the member stiffness matrix $[K_e]$. This latter method for pin behavior simulated implicitly in the member stiffness matrix is called the condensation method, which is described next.

1.12.1 Condensation Method

The rotational freedom for any member can be expressed explicitly outside the domain of the stiffness matrix. In doing so, the rotational freedom is regarded as a variable dependent on other displacement quantities and can be eliminated from the member stiffness matrix. The process of elimination is called *condensation* and hence the name of this method.

In using the condensation method, while the stiffness matrix of the member needs to be modified according to its end connection condition, the internal loads associated with that member also need to be modified. There are three cases that need to be considered for a member. They are (i) pin at end j, (ii) pin at end i, and (iii) pins at both ends.

Case i: Pin at end j

Consider part of a structure shown in Figure 1.28. The freedom codes for member 2 with a pin at end j are $\{1, 2, 3, 4, 5, X\}$ where the rotational freedom X is treated as a dependent variable outside the structure equilibrium equation, leaving the member with only 5 freedom codes pertaining to the structure stiffness matrix $[K]$. Note that the rotational freedom code '6' belongs to member 3.

From Equation (1.28) for member 2 with internal loads,

$$\begin{Bmatrix} N_i \\ Q_i \\ M_i \\ N_j \\ Q_j \\ M_{jX} \end{Bmatrix} = \begin{bmatrix} \frac{EA}{L} & 0 & 0 & -\frac{EA}{L} & 0 & 0 \\ 0 & \frac{12EI}{L^3} & \frac{6EI}{L^2} & 0 & -\frac{12EI}{L^3} & \frac{6EI}{L^2} \\ 0 & \frac{6EI}{L^2} & \frac{4EI}{L} & 0 & -\frac{6EI}{L^2} & \frac{2EI}{L} \\ -\frac{EA}{L} & 0 & 0 & \frac{EA}{L} & 0 & 0 \\ 0 & -\frac{12EI}{L^3} & -\frac{6EI}{L^2} & 0 & \frac{12EI}{L^3} & -\frac{6EI}{L^2} \\ 0 & \frac{6EI}{L^2} & \frac{2EI}{L} & 0 & -\frac{6EI}{L^2} & \frac{4EI}{L} \end{bmatrix} \begin{Bmatrix} u_i \\ v_i \\ \theta_i \\ u_j \\ v_j \\ \theta_{jX} \end{Bmatrix} + \begin{Bmatrix} 0 \\ Q_{Fi} \\ M_{Fi} \\ 0 \\ Q_{Fj} \\ M_{Fj} \end{Bmatrix} \tag{1.29}$$

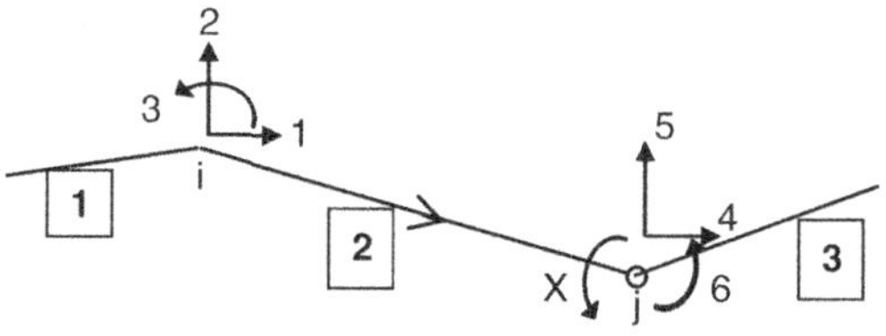

FIGURE 1.28. Member with a pin at end *j*.

where the rotation at end j is θ_{jX} corresponding to a rotational freedom code 'X'. Expanding the last equation in Equation (1.29) and given $M_{jX} = 0$ for a pin, θ_{jX} can be derived as

$$\theta_{jX} = -\frac{3}{2L}v_i - \frac{1}{2}\theta + i\frac{3}{2L}v_j - \frac{M_{Fj}}{4EI/L} \tag{1.30}$$

By substituting Equation (1.30) into the other equations of Equation (1.29), a modified 5 × 5 member stiffness matrix, $[K_{ej}]$, and a modified fixed-end force vector, $\{P_{Fj}\}$, for a member with pin at end j are obtained:

$$\begin{Bmatrix} N_i \\ Q_i \\ M_i \\ N_j \\ Q_j \end{Bmatrix} = [K_{ej}] \begin{Bmatrix} u_i \\ v_i \\ \theta_i \\ u_j \\ v_j \end{Bmatrix} + \{P_{Fj}\} \tag{1.31}$$

where

$$[K_{ej}] = \begin{bmatrix} \frac{EA}{L} & 0 & 0 & -\frac{EA}{L} & 0 \\ 0 & \frac{3EI}{L^3} & \frac{3EI}{L^2} & 0 & -\frac{3EI}{L^3} \\ 0 & \frac{3EI}{L^2} & \frac{3EI}{L} & 0 & -\frac{3EI}{L^2} \\ -\frac{EA}{L} & 0 & 0 & \frac{EA}{L} & 0 \\ 0 & -\frac{3EI}{L^3} & -\frac{3EI}{L^2} & 0 & \frac{3EI}{L^3} \end{bmatrix} \tag{1.32}$$

$$\{P_{Fj}\} = \begin{Bmatrix} 0 \\ Q'_{Fi} \\ M'_{Fi} \\ 0 \\ Q'_{Fj} \end{Bmatrix} = \begin{Bmatrix} 0 \\ Q_{Fi} - \frac{3M_{Fj}}{2L} \\ M_{Fi} - \frac{M_{Fj}}{2} \\ 0 \\ Q_{Fj} + \frac{3M_{Fj}}{2L} \end{Bmatrix} \tag{1.33}$$

Equation (1.33) represents the support reactions equal to those of a propped cantilever beam. Explicit expressions for the coefficients in $\{P_{Fj}\}$ are given in Table 1.2 in Section 1.12.1.4.

The member stiffness matrix in the global coordinate system can be derived as before using a modified transformation matrix, $[T_j]$, which is given as

$$[T_j] = \begin{bmatrix} \cos\alpha & -\sin\alpha & 0 & 0 & 0 \\ \sin\alpha & \cos\alpha & 0 & 0 & 0 \\ 0 & 0 & 1 & 0 & 0 \\ 0 & 0 & 0 & \cos\alpha & -\sin\alpha \\ 0 & 0 & 0 & \sin\alpha & \cos\alpha \end{bmatrix} \tag{1.34}$$

Accordingly, for a member with a pin at end j, the member stiffness matrix in the global coordinate system is

$$\left[K_{ej}^{g}\right] = [T_j]\,[K_{ej}]\,[T_j]^t$$

$$= \begin{bmatrix} C^2\frac{EA}{L}+S^2\frac{3EI}{L^3} & SC\left(\frac{EA}{L}-\frac{3EI}{L^3}\right) & -S\frac{3EI}{L^2} & -\left(C^2\frac{EA}{L}+S^2\frac{3EI}{L^3}\right) & -SC\left(\frac{EA}{L}-\frac{3EI}{L^3}\right) \\ & S^2\frac{EA}{L}+C^2\frac{3EI}{L^3} & C\frac{3EI}{L^2} & -SC\left(\frac{EA}{L}-\frac{3EI}{L^3}\right) & -\left(S^2\frac{EA}{L}+C^2\frac{3EI}{L^3}\right) \\ & & \frac{3EI}{L} & S\frac{3EI}{L^2} & -C\frac{3EI}{L^2} \\ & & & C^2\frac{EA}{L}+S^2\frac{3EI}{L^3} & SC\left(\frac{EA}{L}-\frac{3EI}{L^3}\right) \\ & \textit{Symmetric} & & & S^2\frac{EA}{L}+C^2\frac{3EI}{L^3} \end{bmatrix} \tag{1.35}$$

The modified fixed-end force vector in the global coordinate system, $\{P_{Ej}^{g}\}$, can be derived in a way similar to Equation (1.27).

There are two ways to calculate the member forces. The first way is to use Equation (1.24), for which the end rotation at end j of the member in $\{D_e^g\}$ is replaced by θ_{jX} calculated from Equation (1.30). The second way is to use a form similar to Equation (1.24):

$$\{P\} = [K_{ej}]\,[T_j]^t\{D_e^g\} \tag{1.36a}$$

where, through Equations (1.32) and (1.34),

$$[K_{ej}][T_j]^t = \begin{bmatrix} C\frac{EA}{L} & S\frac{EA}{L} & 0 & -C\frac{EA}{L} & -S\frac{EA}{L} \\ -S\frac{3EI}{L^3} & C\frac{3EI}{L^3} & \frac{3EI}{L^2} & S\frac{3EI}{L^3} & -C\frac{3EI}{L^3} \\ -S\frac{3EI}{L^2} & C\frac{3EI}{L^2} & \frac{3EI}{L} & S\frac{3EI}{L^2} & -C\frac{3EI}{L^2} \\ -C\frac{EA}{L} & -S\frac{EA}{L} & 0 & C\frac{EA}{L} & S\frac{EA}{L} \\ S\frac{3EI}{L^3} & -C\frac{3EI}{L^3} & -\frac{3EI}{L^2} & -S\frac{3EI}{L^3} & C\frac{3EI}{L^3} \end{bmatrix} \quad (1.36b)$$

The 5×1 member displacement vector $\{D_e^g\}$ in Equation (1.36a) is extracted from $\{D\}$ according to the 5 freedom codes $\{1, 2, 3, 4, 5\}$ shown in Figure 1.28 for the member.

Case ii: Pin at end i

This case is shown in Figure 1.29 where member 2 has a pin at end i with an independent rotational freedom code Y. The freedom codes for member 2 with a pin at end i are $\{1, 2, Y, 4, 5, 6\}$. Note that the freedom code 3 belongs to member 1.

By writing

$$\{P\} = \begin{Bmatrix} N_i \\ Q_i \\ M_{iY} \\ N_j \\ Q_j \\ M_j \end{Bmatrix}, \{d\} = \begin{Bmatrix} u_i \\ v_i \\ \theta_{iY} \\ u_j \\ v_j \\ \theta_j \end{Bmatrix}$$

and given $M_{jY} = 0$ for a pin at end i, θ_{iY} can be derived as

$$\theta_{iY} = -\frac{3}{2L}v_i + \frac{3}{2L}v_j - \frac{1}{2}\theta_j - \frac{M_{Fi}}{4EI/L} \quad (1.37)$$

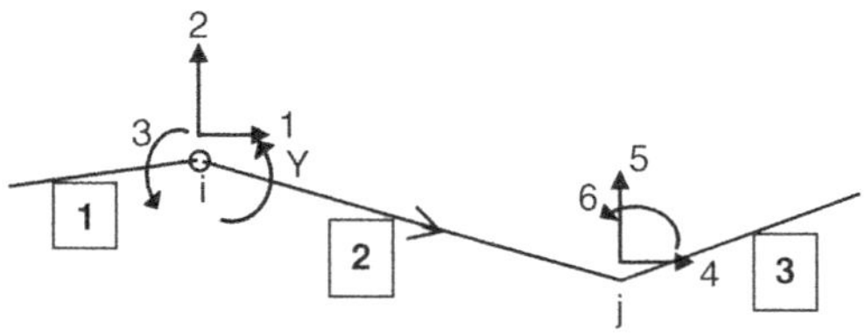

FIGURE 1.29. Member with a pin at end *i*.

The corresponding matrices for this case can be derived in a way similar to Case *i* mentioned earlier. The results are

$$\begin{Bmatrix} N_i \\ Q_i \\ N_j \\ Q_j \\ M_j \end{Bmatrix} = [K_{ei}] \begin{Bmatrix} u_i \\ v_i \\ u_j \\ v_j \\ \theta_j \end{Bmatrix} + (P_{Fi}) \tag{1.38}$$

where

$$[K_{ei}] = \begin{bmatrix} \frac{EA}{L} & 0 & -\frac{EA}{L} & 0 & 0 \\ 0 & \frac{3EI}{L^3} & 0 & -\frac{3EI}{L^3} & \frac{3EI}{L^2} \\ -\frac{EA}{L} & 0 & \frac{EA}{L} & 0 & 0 \\ 0 & -\frac{3EI}{L^3} & 0 & \frac{3EI}{L^3} & -\frac{3EI}{L^2} \\ 0 & \frac{3EI}{L^2} & 0 & -\frac{3EI}{L^2} & \frac{3EI}{L} \end{bmatrix} \tag{1.39}$$

$$\{P_{Fi}\} = \begin{Bmatrix} 0 \\ Q''_{Fi} \\ 0 \\ Q''_{Fj} \\ M''_{Fj} \end{Bmatrix} = \begin{Bmatrix} 0 \\ Q_{Fi} - \frac{3M_{Fi}}{2L} \\ 0 \\ Q_{Fj} + \frac{3M_{Fi}}{2L} \\ M_{Fj} - \frac{M_{Fi}}{2} \end{Bmatrix} \tag{1.40}$$

Equation (1.40) represents the support reactions equal to those of a propped cantilever beam. Explicit expressions for the coefficients in $\{P_{Fi}\}$ are given in Table 1.2 in Section 1.12.1.4.

$$[T_i] = \begin{bmatrix} \cos\alpha & -\sin\alpha & 0 & 0 & 0 \\ \sin\alpha & \cos\alpha & 0 & 0 & 0 \\ 0 & 0 & \cos\alpha & -\sin\alpha & 0 \\ 0 & 0 & \sin\alpha & \cos\alpha & 0 \\ 0 & 0 & 0 & 0 & 1 \end{bmatrix} \tag{1.41}$$

Accordingly, for the member with pin at end *i*, the member stiffness matrix in the global coordinate system is

$$[K_{ei}^{g}] = \begin{bmatrix} C^2\frac{EA}{L}+S^2\frac{3EI}{L^3} & SC\left(\frac{EA}{L}-\frac{3EI}{L^3}\right) & -\left(C^2\frac{EA}{L}+S^2\frac{3EI}{L^3}\right) & -SC\left(\frac{EA}{L}-\frac{3EI}{L^3}\right) & -S\frac{3EI}{L^2} \\ & S^2\frac{EA}{L}+C^2\frac{3EI}{L^3} & -SC\left(\frac{EA}{L}-\frac{3EI}{L^3}\right) & -\left(S^2\frac{EA}{L}+C^2\frac{3EI}{L^3}\right) & C\frac{3EI}{L^2} \\ & & C^2\frac{EA}{L}+S^2\frac{3EI}{L^3} & SC\left(\frac{EA}{L}-\frac{3EI}{L^3}\right) & S\frac{3EI}{L^2} \\ & & & S^2\frac{EA}{L}+C^2\frac{3EI}{L^3} & -C\frac{3EI}{L^2} \\ & \text{Symmetric} & & & \frac{3EI}{L} \end{bmatrix} \tag{1.42}$$

and

$$[K_{ei}][T_i]^t = \begin{bmatrix} C\frac{EA}{L} & S\frac{EA}{L} & -C\frac{EA}{L} & -S\frac{EA}{L} & 0 \\ -S\frac{3EI}{L^3} & C\frac{3EI}{L^3} & S\frac{3EI}{L^3} & -C\frac{3EI}{L^3} & \frac{3EI}{L^2} \\ -C\frac{EA}{L} & -S\frac{EA}{L} & C\frac{EA}{L} & S\frac{EA}{L} & 0 \\ S\frac{3EI}{L^3} & -C\frac{3EI}{L^3} & -S\frac{3EI}{L^3} & C\frac{3EI}{L^3} & -\frac{3EI}{L^2} \\ -S\frac{3EI}{L^2} & C\frac{3EI}{L^2} & S\frac{3EI}{L^2} & -C\frac{3EI}{L^2} & \frac{3EI}{L} \end{bmatrix} \tag{1.43}$$

The 5 × 1 member displacement vector $\{D_e^g\}$ is extracted from $\{D\}$ according to the 5 freedom codes $\{1, 2, 4, 5, 6\}$ for the member.

Case iii: Pins at both ends i *and* j

This case is shown in Figure 1.30 where member 2 has a pin at both ends *i* and *j*. The freedom codes for member 2 are $\{1, 2, Y, 4, 5, X\}$. Note that the freedom codes 3 and 6 belong to members 1 and 3, respectively.

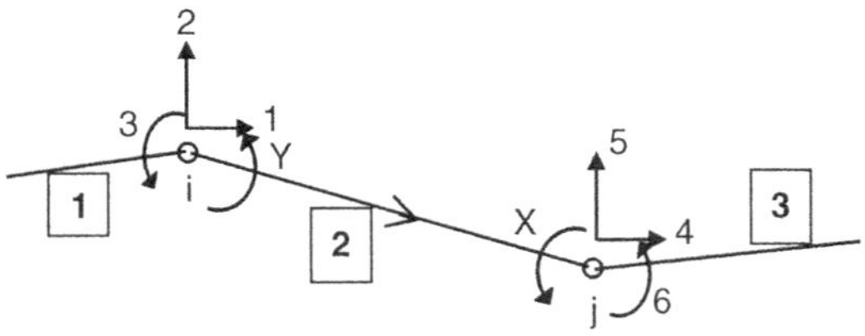

FIGURE 1.30. Member with a pin at end *i*.

In this case, substitute $M_{iY} = M_{jX} = 0$ into Equation (1.29), we obtain

$$\theta_{iY} = \frac{v_j - v_i}{L} + \frac{M_{Fj} - 2M_{Fi}}{6EI/L} \tag{1.44a}$$

$$\theta_{jX} = \frac{v_j - v_i}{L} + \frac{M_{Fi} - 2M_{Fj}}{6EI/L} \tag{1.44b}$$

and

$$\begin{Bmatrix} N_i \\ Q_i \\ N_j \\ Q_j \end{Bmatrix} = [K_{eij}] \begin{Bmatrix} u_i \\ v_i \\ u_j \\ v_j \end{Bmatrix} + \{P_{Fij}\} \tag{1.45}$$

where

$$\{K_{eij}\} = \frac{EA}{L} \begin{bmatrix} 1 & 0 & -1 & 0 \\ 0 & 0 & 0 & 0 \\ -1 & 0 & 1 & 0 \\ 0 & 0 & 0 & 0 \end{bmatrix} \tag{1.46}$$

$$\{P_{Fij}\} = \begin{Bmatrix} 0 \\ Q'''_{Fi} \\ 0 \\ Q'''_{Fj} \end{Bmatrix} = \begin{Bmatrix} 0 \\ Q_{Fi} - \dfrac{(M_{Fi} + M_{Fj})}{L} \\ 0 \\ Q_{Fj} + \dfrac{(M_{Fi} + M_{Fj})}{L} \end{Bmatrix} \tag{1.47}$$

Equation (1.47) represents the support reactions equal to those of a simply supported beam. Explicit expressions for the coefficients in $\{P_{Fij}\}$ are given in Table 1.2 in Section 1.12.1.4.

It is noted that $[K_{eij}]$ is in fact the stiffness matrix of a truss member. The transformation matrix for the member in this case is

$$[T_{ij}] = \begin{bmatrix} C & -S & 0 & 0 \\ S & C & 0 & 0 \\ 0 & 0 & C & -S \\ 0 & 0 & S & C \end{bmatrix} \tag{1.48}$$

The corresponding stiffness matrix in the global coordinate system for a member with pins at both ends is

$$[K^g_{eij}] = \frac{EA}{L} \begin{bmatrix} C^2 & CS & -C^2 & -CS \\ & S^2 & -CS & -S^2 \\ & & C^2 & CS \\ & \textit{Symmetric} & & S^2 \end{bmatrix} \tag{1.49}$$

and

$$\left[K^g_{eij}\right]\left[T_{ij}\right]^t = \frac{EA}{L}\begin{bmatrix} C & S & -C & -S \\ 0 & 0 & 0 & 0 \\ -C & -S & C & S \\ 0 & 0 & 0 & 0 \end{bmatrix} \tag{1.50}$$

Modified Fixed-End Force Vector

The explicit expressions for the coefficients of the modified fixed-end force vectors given in Equations (1.33), (1.40), and (1.47) are summarized in Table 1.2.

TABLE 1.2
Modified fixed-end forces for members with pins

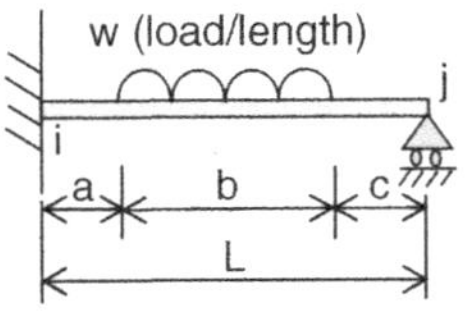	Q'_{Fi}	$= \dfrac{wb(\frac{b}{2}+c)}{L} + \dfrac{M'_{Fi}}{L}$
	Q'_{Fj}	$= wb - Q'_{Fi}$
	M'_{Fi}	$= \dfrac{w}{8L^2}\left[(b+2c)b\left(2L^2 - c^2 - (b+c)^2\right)\right]$
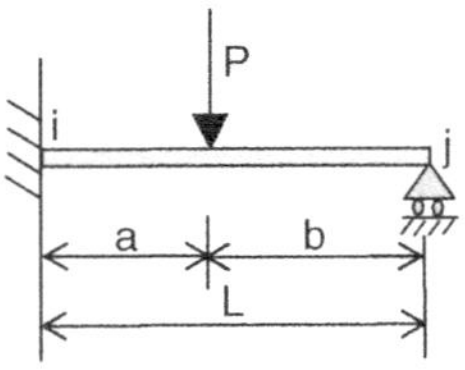	Q''_{Fi}	$= P - Q''_{Fj}$
	Q''_{Fj}	$= \dfrac{Pa^2}{2L^3}(b+2L)$
	M''_{Fi}	$= \dfrac{Pb(L^2-b^2)}{2L^2}$
	Moment under load	$= \dfrac{Pb}{2}\left(2 - \dfrac{3b}{L} + \dfrac{b^3}{L^3}\right)$
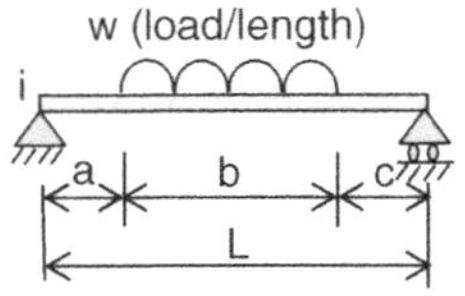	Q'''_{Fi}	$= \dfrac{wb}{L}\left(\dfrac{b}{2} + c\right)$
	Q'''_{Fj}	$= \dfrac{wb}{L}\left(\dfrac{b}{2} + a\right)$
	M_{max}	$= w\left(\dfrac{x^2 - a^2}{2}\right)$ at $x = a + \dfrac{Q'''_{Fi}}{w}$
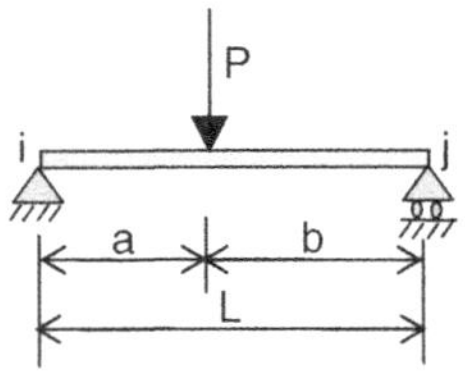	Q'''_{Fi}	$= \dfrac{Pb}{L}$
	Q'''_{Fj}	$= \dfrac{Pa}{L}$
	Moment under load	$= \dfrac{Pab}{L}$

Procedure for Using Condensation Method

1. For any joints with pins, determine whether the connecting members have (a) no pin, (b) pin at end i, (c) pin at end j, or (d) pins at both ends.
2. Use the appropriate stiffness matrix for the cases just given for all members.
3. Assign freedom codes to each joint.
4. Assemble the structure stiffness matrix $[K]$.
5. After solving the structure equilibrium equation, calculate the angle of rotation θ_{jX} or θ_{iY} for each pin using Equations (1.30), (1.37), or (1.44). Calculate the member forces accordingly.

1.12.2 Methods to Model Pin

There are a number of ways to model a joint with a pin using the formulations given in the previous section. Consider a pinned joint connecting two members 1 and 2. There are four ways of formulation for use in the stiffness method of analysis as shown in Figure 1.31. Figure 1.31a is based on the extra freedom method where both members 1 and 2 have independent rotations D_3 and D_4 using the full 6×6 member stiffness matrix. Figure 1.31b is based on the condensation method for member 1 using the formulation for pin at end j as given in Section 1.12.1.1, whereas member 2 retains use of the full 6×6 member stiffness matrix. Figure 1.31c is also based on the condensation method for member 2 using the formulation for pin at end i

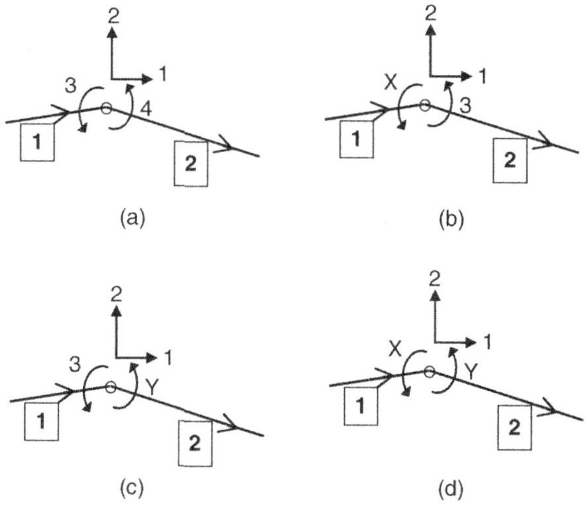

FIGURE 1.31. Modeling pin at a joint.

as given in Section 1.12.1.2, whereas member 1 retains use of the full 6×6 member stiffness matrix. Figure 1.31d is based on the condensation method using the formulation for pin at end j for member 1 and pin at end i for member 2.

Example 1.5 Determine the displacements and forces in the beam ABC with a pin at B shown in Figure 1.32. Ignore the effect of axial force. $E = 2000$ kN/m^2, $I = 0.015$ m^4.

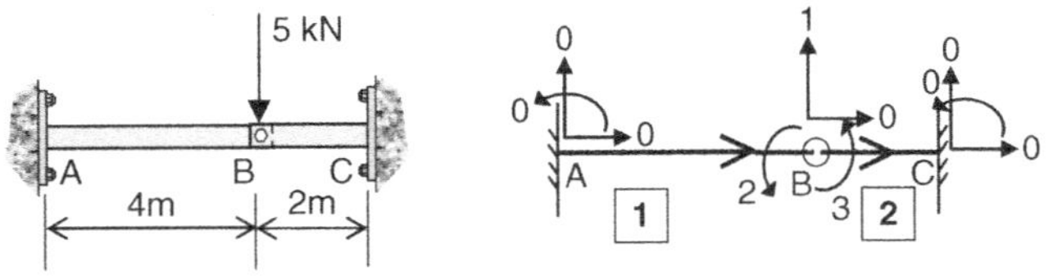

(a) Beam with pin at B

(b) Freedom codes–Extra Freedom Method

FIGURE 1.32. Example 1.5.

Solution

(i) Extra Freedom Method

When the axial force effect is ignored, a zero freedom code is assigned to the axial deformation of the members. Thus, the structure has a total of 3 degrees of freedom shown in Figure 1.32b.

For all matrices, only the coefficients corresponding to nonzero freedom codes will be shown.

For member 1,

$$[K_e^g]_1 = \begin{array}{c} \begin{array}{cccccc} 0 & 0 & 0 & 0 & 1 & 2 \end{array} \\ \left[\begin{array}{cccccc} .. & .. & .. & .. & .. & .. \\ & .. & .. & .. & .. & .. \\ & & .. & .. & .. & .. \\ & & & .. & .. & .. \\ & \textit{Symmetric} & & & 5.625 & -11.25 \\ & & & & & 30 \end{array}\right] \end{array} \begin{array}{c} \\ 0 \\ 0 \\ 0 \\ 0 \\ 1 \\ 2 \end{array}$$

For member 2,

$$[K_e^g]_2 = \begin{array}{c} \begin{array}{cccccc} 0 & 1 & 3 & 0 & 0 & 0 \end{array} \\ \left[\begin{array}{cccccc} .. & .. & .. & .. & .. & .. \\ & 45 & 45 & .. & .. & .. \\ & & 60 & .. & .. & .. \\ & & & .. & .. & .. \\ & \textit{Symmetric} & & & .. & .. \\ & & & & & .. \end{array}\right] \end{array} \begin{array}{c} \\ 0 \\ 1 \\ 3 \\ 0 \\ 0 \\ 0 \end{array}$$

Hence,

$$[K] = \begin{bmatrix} 50.625 & -11.25 & 45 \\ -11.25 & 30 & 0 \\ 45 & 0 & 60 \end{bmatrix}$$

$$\{F\} = \begin{Bmatrix} -5 \\ 0 \\ 0 \end{Bmatrix}$$

By solving the structure equilibrium equation $\{F\} = [K]\{D\}$ for $\{D\}$, we obtain

$$\{D\} = \begin{Bmatrix} D_1 \\ D_2 \\ D_3 \end{Bmatrix} = \begin{Bmatrix} -0.395 \\ -0.148 \\ 0.296 \end{Bmatrix}$$

Figure 1.33 shows the deflection and rotations of the members.

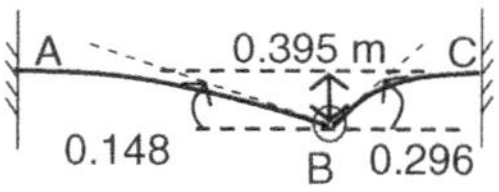

FIGURE 1.33. Deflection and rotations.

The member forces for member 1 are

$$\{P\}_1 = \begin{bmatrix} .. & .. & .. & .. & 0 & 0 \\ .. & .. & .. & .. & -5.625 & 11.25 \\ .. & .. & .. & .. & -11.25 & 15 \\ .. & .. & .. & .. & 0 & 0 \\ .. & .. & .. & .. & 5.625 & -11.25 \\ .. & .. & .. & .. & -11.25 & 30 \end{bmatrix} \begin{Bmatrix} 0 \\ 0 \\ 0 \\ 0 \\ -0.395 \\ -0.148 \end{Bmatrix} = \begin{Bmatrix} 0 \\ 0.556 \\ 2.223 \\ 0 \\ -0.556 \\ 0 \end{Bmatrix}$$

The member forces for member 2 are

$$\{P\}_2 = \begin{bmatrix} .. & 0 & 0 & .. & .. & .. \\ .. & 45 & 45 & .. & .. & .. \\ .. & 45 & 60 & .. & .. & .. \\ .. & 0 & 0 & .. & .. & .. \\ .. & -45 & -45 & .. & .. & .. \\ .. & 45 & 30 & .. & .. & .. \end{bmatrix} \begin{Bmatrix} 0 \\ -0.395 \\ 0.296 \\ 0 \\ 0 \\ 0 \end{Bmatrix} = \begin{Bmatrix} 0 \\ -4.444 \\ 0 \\ 0 \\ 4.444 \\ -8.891 \end{Bmatrix}$$

The member forces for the structure are shown in Figure 1.34.

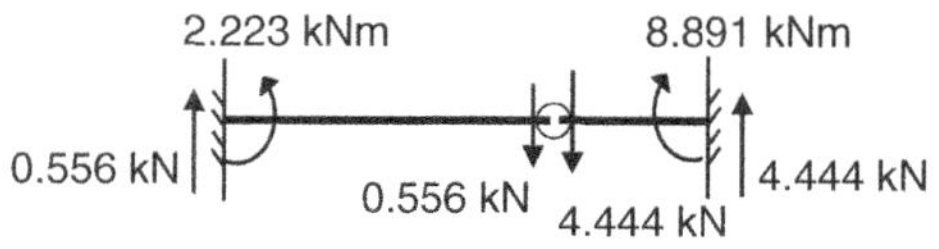

FIGURE 1.34. Member forces.

(ii) Method of Condensation

In using this method, the stiffness matrix of member 1 is condensed so that rotation at end j, denoted as X in Figure 1.35, becomes a dependent variable. The freedom codes of the structure are also shown in Figure 1.35.

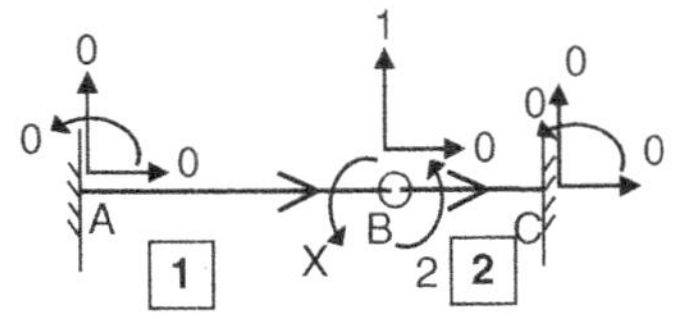

FIGURE 1.35. Freedom codes: method of condensation.

The stiffness matrix of member 1 is given by Equation (1.32) as

$$[K_{ej}]_1 = \left[K_{ej}^g\right]_1 = \begin{array}{c} \begin{array}{ccccc} 0 & 0 & 0 & 0 & 1 \end{array} \\ \left[\begin{array}{ccccc} .. & .. & .. & .. & .. \\ & .. & .. & .. & .. \\ & & .. & .. & .. \\ & \textit{Symmetric} & & & .. \\ & & & & 1.4063 \end{array}\right] \end{array} \begin{array}{c} \\ 0 \\ 0 \\ 0 \\ 0 \\ 1 \end{array}$$

For member 2,

$$[K_e^g]_2 = \begin{array}{c} \begin{array}{cccccc} 0 & 1 & 2 & 0 & 0 & 0 \end{array} \\ \left[\begin{array}{cccccc} .. & .. & .. & .. & .. & .. \\ & 45 & 45 & .. & .. & .. \\ & & 60 & .. & .. & .. \\ & & & .. & .. & .. \\ & \textit{Symmetric} & & & .. & .. \\ & & & & & .. \end{array}\right] \end{array} \begin{array}{c} \\ 0 \\ 1 \\ 2 \\ 0 \\ 0 \\ 0 \end{array}$$

Hence, the structure stiffness matrix, of size 2×2, can be assembled as

$$[K] = \begin{bmatrix} 46.406 & 45 \\ 45 & 60 \end{bmatrix}$$

The load vector is

$$\{F\} = \begin{Bmatrix} -5 \\ 0 \end{Bmatrix}$$

By solving $\{F\} = [K]\{D\}$ for $\{D\}$, we obtain

$$\{D\} = \begin{Bmatrix} D_1 \\ D_2 \end{Bmatrix} = \begin{Bmatrix} -0.395 \\ 0.296 \end{Bmatrix}$$

The rotation θ_{jX} for member 1 can be obtained from Equation (1.30) as

$$\theta_{jX} = -\frac{3}{2L}v_i - \frac{1}{2}\theta_i + \frac{3}{2L}v_j = \frac{3}{2 \times 4}(-0.395) = -0.148$$

The member forces for member 2 can be calculated using Equation (1.36a)

$$\{P\}_1 = [K_{ej}][T_j]^t\{D_e^g\}$$

$$= \begin{bmatrix} .. & .. & .. & .. & 0 \\ .. & .. & .. & .. & -1.406 \\ .. & .. & .. & .. & -5.625 \\ .. & .. & .. & .. & 0 \\ . & .. & .. & .. & 1.406 \end{bmatrix} \begin{Bmatrix} 0 \\ 0 \\ 0 \\ 0 \\ -0.395 \end{Bmatrix} = \begin{Bmatrix} 0 \\ 0.556 \\ 2.223 \\ 0 \\ -0.556 \end{Bmatrix}$$

which are the same as those calculated before.

1.13 Temperature Effects

Most materials expand when subject to temperature rise. For a steel member in a structure, the expansion due to temperature rise is restrained by the other members connected to it. The restraint imposed on the heated member generates internal member forces exerted on the structure. For uniform temperature rise in a member, the internal member forces are axial and compressive, and their effects can be treated in the same way as for internal loads described in Section 1.11.

1.13.1 Uniform Temperature

The fixed-end force vector $\{P_F\}$ for a steel member shown in Figure 1.36 subject to a temperature rise of $(T - T_o)$, where T is the current temperature and T_o is the ambient temperature of the member, is given by

$$\{P_F\} = \begin{Bmatrix} N_{Fi} \\ 0 \\ 0 \\ N_{Fj} \\ 0 \\ 0 \end{Bmatrix} \tag{1.51}$$

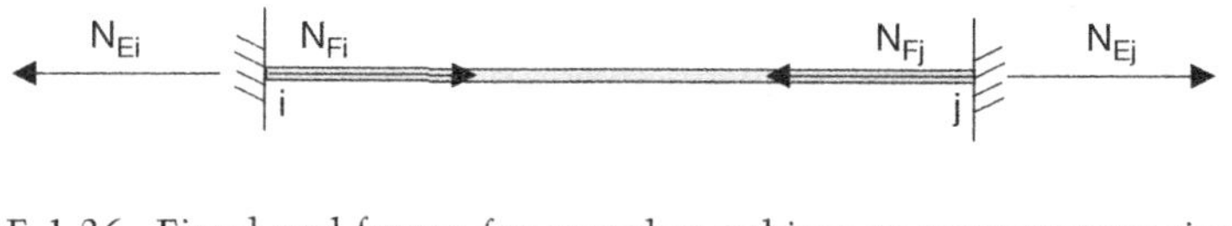

FIGURE 1.36. Fixed-end forces for member subject to temperature rise.

where

$$N_{Fi} = -N_{Fj} = E_T A\alpha(T - T_o) \tag{1.52}$$

E_T = modulus of elasticity at temperature T,
A = cross-sectional area,
α = coefficient of linear expansion.

As before, the equivalent force vector is $\{P_E\} = -\{P_F\}$.

In Equation (1.52), E_T is often treated as a constant for low temperature rise. However, under extreme loading conditions, such as steel in a fire, the value of E_T deteriorates significantly over a range of temperatures. The deterioration rate of steel at elevated temperature is often expressed as a ratio of E_T/E_o. This ratio has many forms according to the design codes adopted by different countries. In Australia and America, the ratio of E_T/E_o is usually expressed as

$$\frac{E_T}{E_o} = 1.0 + \frac{T}{2000 \ln\left[\dfrac{T}{1100}\right]} \quad \text{for } 0^\circ C < T \leq 600^\circ C$$

$$= \frac{690\left(1 - \dfrac{T}{1000}\right)}{T - 53.5} \quad \text{for } 600^\circ C < T \leq 1000^\circ C \tag{1.53a}$$

In Europe, the ratio of E_T/E_o, given in tabulated form in the Eurocode, can be approximated as

$$\frac{E_T}{E_o} = 1 - e^{-9.7265 \times 0.9947^T} \tag{1.53b}$$

Although the coefficient of linear expansion α also varies with temperature for steel, its variation is insignificantly small. Therefore, a constant value is usually adopted. The overall effect of rising temperature and deteriorating stiffness for a steel member is that the fixed-end compressive force increases initially up to a peak at about 500°C, beyond which the compressive force starts to decrease. The variation of the fixed-end compressive force, expressed as a dimensionless ratio relative to its value at 100°C using a varying modulus of elasticity according to Equation (1.53a), is shown in Figure 1.37. For comparison purpose, the variation of the fixed-end compressive force using a constant value of E_T is also shown in Figure 1.37.

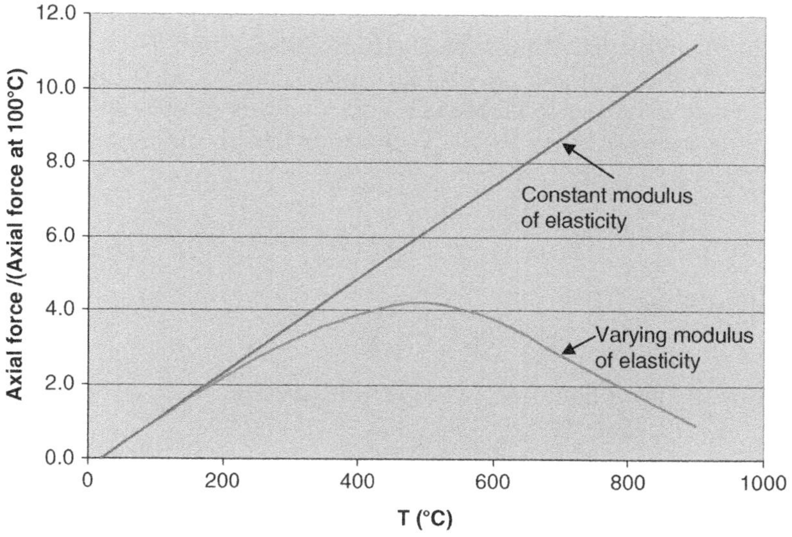

FIGURE 1.37. Variation of fixed-end compressive force with temperature.

1.13.2 Temperature Gradient

For a member subject to a linearly varying temperature across its cross section with T_t = temperature at the top of the cross section and T_b = temperature at the bottom of the cross section, the fixed-end force vector is given by

$$\{P_F\} = \begin{Bmatrix} N_{Fi} \\ 0 \\ M_{Fi} \\ N_{Fj} \\ 0 \\ M_{Fj} \end{Bmatrix} \tag{1.54}$$

where

$$N_{Fi} = -N_{Fj} = \int_A \sigma dA \tag{1.55a}$$

$$\sigma = E_T \alpha (T - T_o) \tag{1.55b}$$

In Equation (1.55a), the integration is carried out for the whole cross section of area A. The stress σ at a point in the cross section corresponds to a temperature T at that point. In practice, integration is approximated by dividing the cross section into a number of horizontal strips, each of which is assumed to have a uniform temperature.

Consider a member of length L with a linearly varying temperature in its cross section subject to an axial force N. If the cross section of the member is divided into n strips and the force in strip i with cross-sectional area A_i and modulus of elasticity E_i is N_i, then, for compatibility with a common axial deformation u for all strips,

$$u = \frac{N_1 L}{E_1 A_1} = \ldots = \frac{N_i L}{E_i A_i} = \ldots = \frac{N_n L}{E_n A_n} \tag{1.56}$$

For equilibrium,

$$N = N_1 + \ldots + N_i + \ldots + N_n \tag{1.57}$$

Substituting Equation (1.56) into Equation (1.57), we obtain

$$N = \frac{\sum_{1}^{n} E_i A_i}{L} u \tag{1.58}$$

By comparing Equation (1.58) with Equation (1.9), it can be seen that for a member with a linearly varying temperature across its cross section,

$$K_{11} = -K_{14} = \frac{\sum_{1}^{n} E_i A_i}{L} \tag{1.59}$$

Equation (1.59) can be rewritten as

$$K_{11} = -K_{14} = \frac{E_o \sum_{1}^{n} m_i A_i}{L} \tag{1.60}$$

in which

$$m_i = \frac{E_i}{E_o} \tag{1.61}$$

The value of m_i can be obtained from Equations (1.53a) or (1.53b). The use of Equation (1.60) is based on the transformed section method, whereby the width of each strip in the cross section is adjusted by multiplying the original width by m_i and the total area is calculated according to the transformed section.

The stiffness coefficients for bending involving EI can also be obtained using the transformed section method. The curvature of the member as a result of bowing due to the temperature gradient across the depth of the cross section is given as

$$\kappa = \alpha \frac{(T_t - T_b)}{d} \tag{1.62}$$

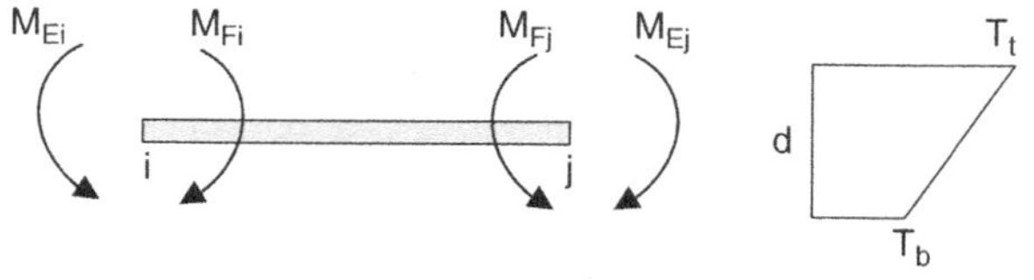

FIGURE 1.38. Fixed-end moments under temperature gradient.

Hence, the fixed-end moments at the ends of the member, as shown in Figure 1.38, are

$$M_{Fi} = -M_{Fj} = -E_a I \alpha \frac{(T_t - T_b)}{d} \tag{1.63}$$

where d is depth of cross section.

Similar to the calculation of the axial stiffness coefficient in Equation (1.60), $E_a I$ in Equation (1.63) is calculated numerically by dividing the cross section into a number of horizontal strips, each of which is assumed to have a uniform temperature. The width of each strip in the cross section is adjusted by multiplying the original width by m_i so that

$$E_a I = \sum_{1}^{n} E_i I_i = E_o \sum_{1}^{n} m_i I_i \tag{1.64}$$

where I_i is calculated about the centroid of the transformed section.

Example 1.6 Determine the axial stiffness EA and bending stiffness EI for the I section shown in Figure 1.39. The section is subject to a linearly varying temperature of 240°C at the top and 600°C at the bottom. Use the European curve [Equation (1.53b)] for the deterioration rate of

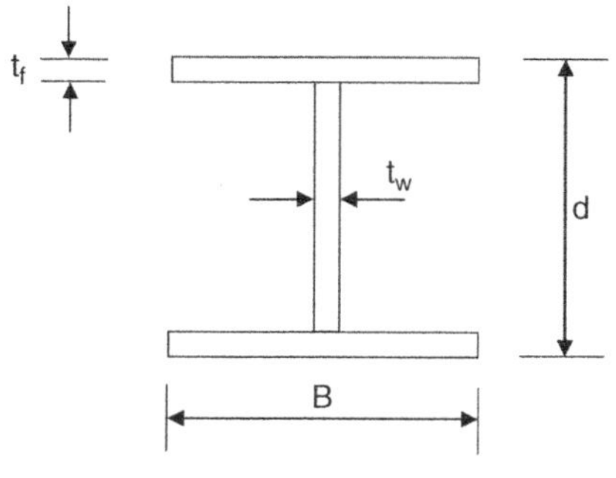

FIGURE 1.39. Example 1.6.

the modulus of elasticity. E_o at ambient temperature = 210,000 MPa. A = 7135 mm^2, I = 158202611 mm^4, B = 172.1 mm, d = 358.6 mm, t_w = 8 mm, t_f = 13 mm.

Solution. The section is divided into 24 strips, 4 in each of the flanges and 16 in the web. The temperature at each strip is taken as the temperature at its centroid. The area of each strip is transformed by multiplying its width by E_T/E_o.

The total area of the transformed section = 4549.9 mm^2.
The centroid of the transformed section from the bottom edge = 239.0 mm.
Total EA for the section = 210000 × 4549.9 = 9.555 × 10^8 N.
The second moment of area of the transformed section = 8.422 × 10^7 mm^4.
Total EI for the section = 210000 × 8.422 × 10^7 = 1.769 × 10^{13} N mm^2.

Problems

1.1. Determine the degree of indeterminacy for the beam shown in Figure P1.1.

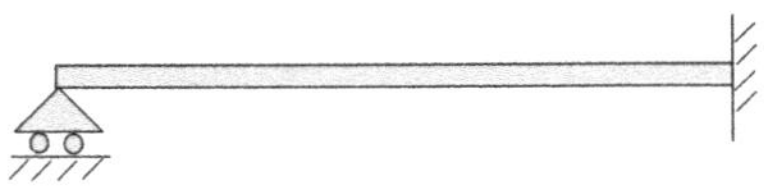

FIGURE P1.1. Problem 1.1.

1.2. Determine the degree of indeterminacy for the beam shown in Figure P1.2.

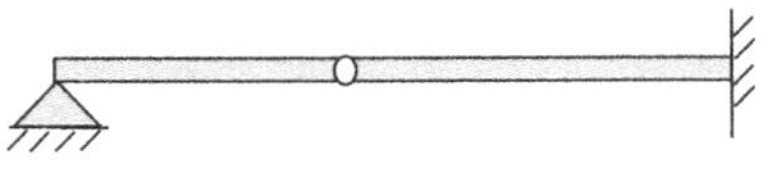

FIGURE P1.2. Problem 1.2.

1.3. Determine the degree of indeterminacy for the continuous beam shown in Figure P1.3.

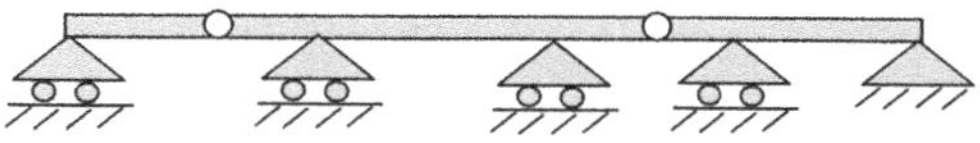

FIGURE P1.3. Problem 1.3.

1.4. Determine the degree of indeterminacy for the frame shown in Figure P1.4.

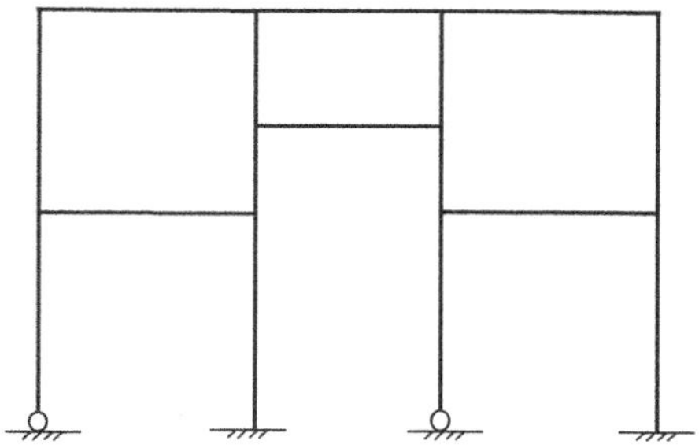

FIGURE P1.4. Problem 1.4.

1.5. The structure ABC shown in Figure P1.5 is subject to a clockwise moment of 5 kNm applied at B. Determine the angles of rotation at A and B using

1. Extra freedom method
2. Condensation method

Ignore axial force effect. $EI = 30$ kNm^2.

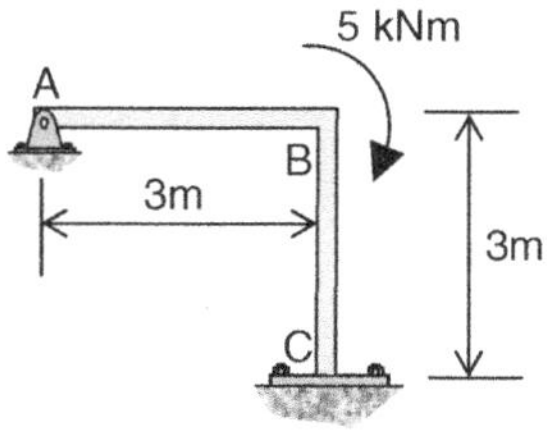

FIGURE P1.5. Problem 1.5.

1.6. The structure shown in Figure P1.6 is fixed at A and C and pinned at B and subject to an inclined force of 300 kN. Determine the forces in the structure and plot the bending moment and shear force diagrams. $E = 2 \times 10^8$ kN/m^2, $I = 1.5 \times 10^{-5}$ m^4, $A = 0.002$ m^2.

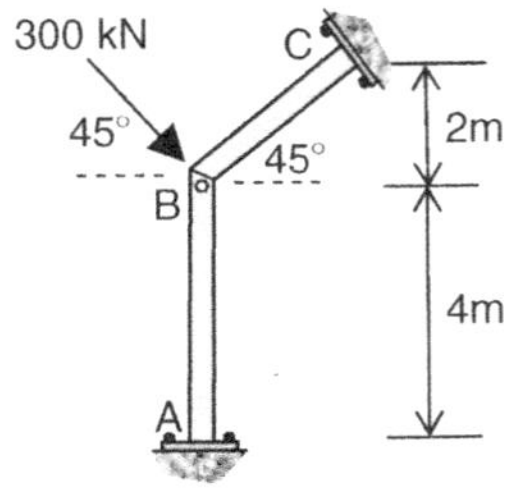

FIGURE P1.6. Problem 1.6.

1.7. The frame ABC shown in Figure P1.7 is pinned at A and fixed to a roller at C. A bending moment of 100 kNm is applied at B. Plot the bending moment and shear force diagrams for the frame. Ignore the effect of axial force in the members. $E = 210000$ kN/m^2, $I = 0.001$ m^4.

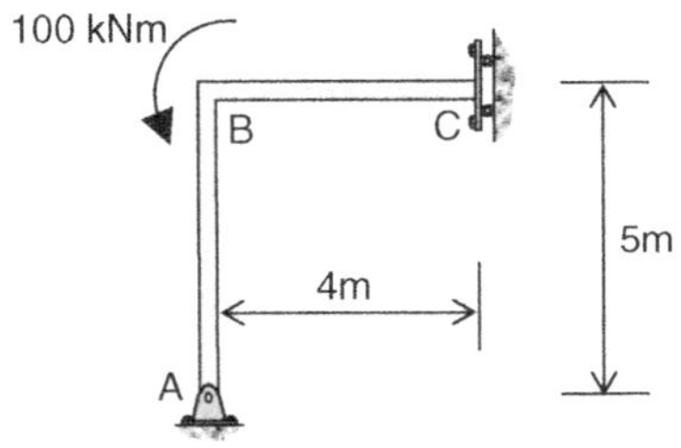

FIGURE P1.7. Problem 1.7.

1.8. A beam ABC shown in Figure P1.8 is pinned at A and fixed at C. A vertical force of 5 kN is applied at B. Determine the displacements of the structure and plot the bending moment and shear force diagrams. Ignore axial force effect. $E = 2000$ kN/m^2, $I = 0.015$ m^4.

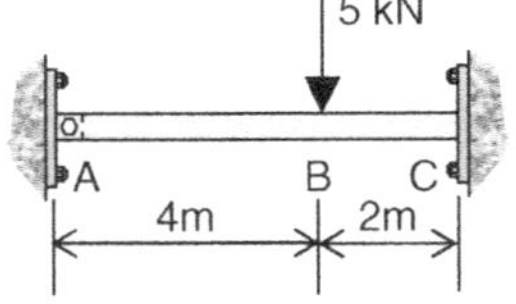

FIGURE P1.8. Example 1.8

1.9. Use the stiffness method to calculate the member forces in the structure shown in Figure P1.9. $E = 2 \times 10^5$ N/mm^2, $A = 6000$ mm^2, $I = 2 \times 10^7$ mm^4.

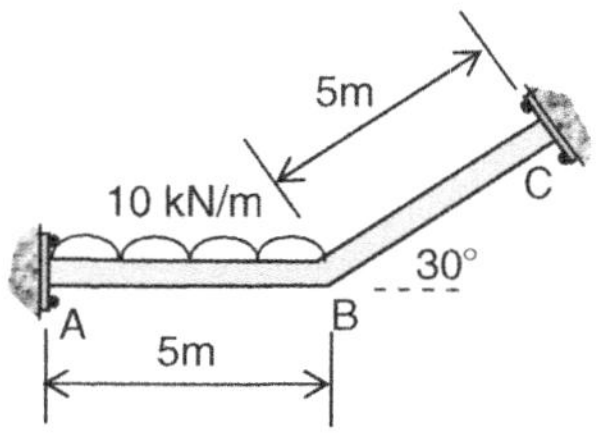

FIGURE P1.9. Problem 1.9.

1.10. Plot the shear force and bending moment diagrams for the continuous beam shown in Figure P1.10. Ignore axial force effect. $E = 3 \times 10^5$ N/mm^2, $I = 2 \times 10^7$ mm^4.

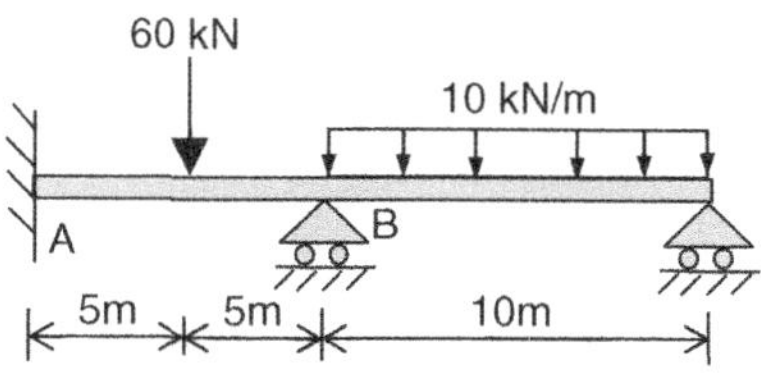

FIGURE P1.10. Problem 1.10.

1.11. Determine the axial stiffness EA and bending stiffness EI for the I section shown in Figure 1.39. The section is subject to a linearly varying temperature of 150°C at the top and 400°C at the bottom. Use the European curve [Equation (1.53b)] for the deterioration rate of the modulus of elasticity. E_o at ambient temperature = 210000 MPa. A = 7135 mm^2, I = 158202611 mm^4. B = 172.1 mm, d = 358.6 mm, t_w = 8 mm, t_f = 13 mm.

Bibliography

1. Wang, C. K. (1963). General computer program for limit analysis. *Proceedings ASCE,* **89**(ST6).
2. Jennings, A., and Majid, K. (1965). An elastic-plastic analysis by computer for framed structures loaded up to collapse. *The Structural Engineer,* **43**(12).
3. Davies, J. M. (1967). Collapse and shakedown loads of plane frames. *Proc. ASCE, J. St. Div.,* **ST3**, pp. 35–50.

4. Chen, W. F., and Sohal, I. (1995). *Plastic design and second-order analysis of steel frames*. New York: Springer-Verlag.
5. Samuelsson, A., and Zienkiewicz, O. C. (2006). History of the stiffness method. *Int. J. Num. Methods in Eng.*, **67**, pp. 149–157.
6. Rangasami, K. S., and Mallick, S. K. (1966). Degrees of freedom of plane and space frames. *The Structural Engineer*, **44**(3), pp. 109–111.

CHAPTER 2

Plastic Behavior of Structures

2.1 Introduction

The early development of plasticity problems in a general finite element approach can be attributed to, for example, Marcal and King,[1] Yamada and Yoshimura,[2] and Zienkiewicz *et al.*[3] The approach has also been adopted by Ueda *et al.*[4] for framed structure applications. The approach is based on plastic flow theory with due consideration given to the plasticity conditions of the elements. A similar formulation was also adopted by Nigam[5] for dynamic analysis. Unlike the elastic approach described in Chapter 1, all the work is based on matrix formulation using stiffness methods for analysis extended to the inelastic range. The complete description of the behavior of a structure from its elastic to plastic state is termed elastoplastic analysis. Other methods making use of the stiffness approach to solving plasticity problems for framed structures include those by Livesley,[6] Davies,[7] and Majid,[8] whose work was mainly on yielding only by pure bending.

Mathematical programming methods have become an important area of research in engineering plasticity in recent years. The general methods of formulation and solution using this approach are typically referenced by Franchi and Cohn,[9] Maier and Munro,[10] and Tin-Loi and Pang.[11] A detailed description of mathematical programming methods is given in Chapter 6.

2.2 Elastic and Plastic Behavior of Steel

This section first describes the structural behavior of a cross section from its elastic state to a fully plastic state under increasing load. The general elastoplastic behavior of a structure will then be given and its application to plastic design method, under certain limitations, is compared with the elastic design method.

Most structural materials undergo an elastic state before a plastic state is reached. This applies to both material behavior of a cross section and the structure as a whole. For a simply supported steel beam with a cross section symmetrical about a horizontal axis under an increasing load applied at midspan, the general stress and strain variations in the cross section at midspan from a fully elastic state to fracture are shown in Figure 2.1. The beam is initially loaded producing an elastic stress $f = f_e$ corresponding to an elastic strain $\varepsilon = \varepsilon_e$ for loading between points A and B shown in Figure 2.1. When point B is reached, the maximum stress in the top and bottom fibers of the cross section becomes yielded such that $f = f_y$, corresponds to a yield strain $\varepsilon = \varepsilon_y$. As the load is increased further, the cross section undergoes a plastification process in which the yielded area becomes larger and larger, spreading inward toward the center of the cross section. This plastification with a relatively constant yield stress f_y occurs between B and C, at which the stress corresponding to strain ε_s starts to increase again. From point C, the cross section enters into a strain-hardening stage until an ultimate stress f_u at D is reached. From point D, the stress starts to decrease with increasing strain until the material fractures at point E. The plastification process is important for steel in plastic design as it ensures that the material has adequate ductility for the cross section to sustain loading beyond its elastic limit at B.

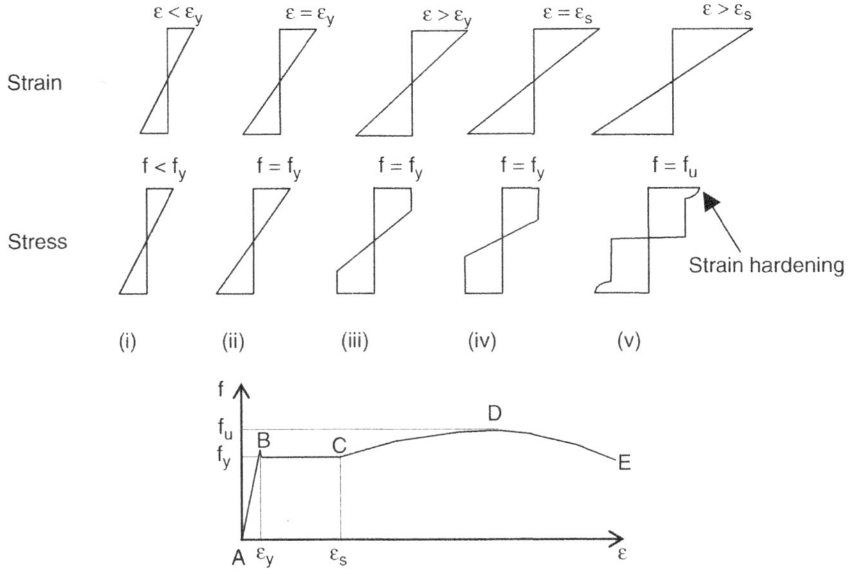

FIGURE 2.1. Stress–strain behavior of a cross section.

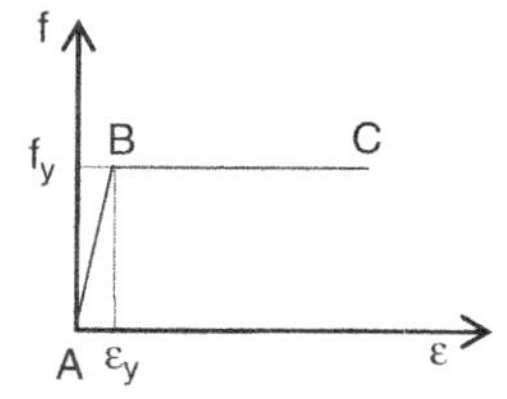

FIGURE 2.2. Elastic perfectly plastic behavior for steel.

For design purposes, it is prudent to ignore the extra strength provided by strain hardening, which becomes smaller in magnitude as the grade strength of steel becomes greater. Hence, for simplicity, steel is always idealized as an elastic-perfectly plastic material with a stress–strain relationship shown in Figure 2.2 and the corresponding cross-section plastification of a symmetric section in Figure 2.3. In Figure 2.2, the stress–strain relationship for the elastic part AB is linear and its slope is equal to the modulus of elasticity.

According to the idealized stress–strain relationship, the continuous plastification of a cross section shown in Figure 2.3 under increasing loading induces continuous increase in bending moment of the cross section. When the extreme fibers of the cross section reach the yield strain, ε_y, with a yield stress, f_y, a yield moment M_y corresponding to a yield curvature κ_y (see Section 2.3 for the definition of curvature) at point B, a moment–curvature relationship shown in Figure 2.4 exists in the section. A further increase in loading causes partial plastification in the cross section, which signals the start of

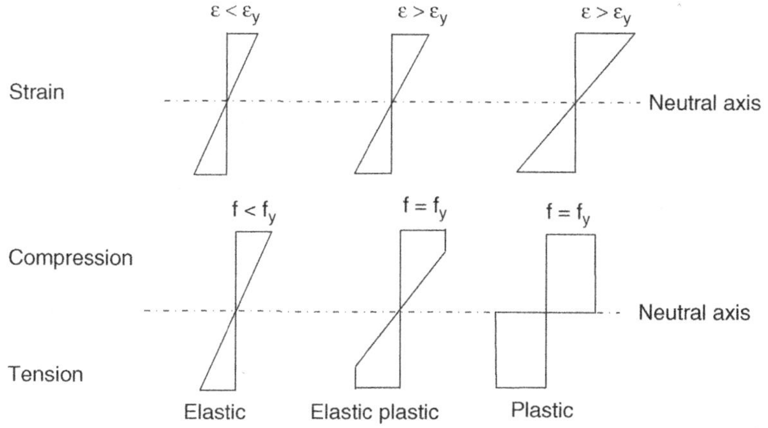

FIGURE 2.3. Plastification of a cross section.

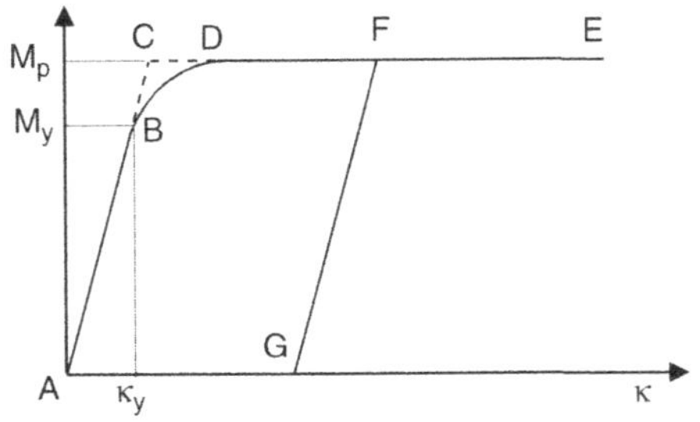

FIGURE 2.4. Moment–curvature relationship of a cross section.

its elastic–plastic state. This elastic–plastic state corresponds to an increase in bending moment from B to D shown in Figure 2.4. When the cross section becomes fully plastic at point D, the maximum moment capacity, called plastic moment M_p, is reached. A further increase in loading increases the strains and hence the curvature in the cross section, whereas the plastic moment remains unchanged. It should be noted that the line DE in Figure 2.4 is not truly horizontal and the point D is difficult to define for most cross-sectional shapes. Unless deflection is a prime consideration, for design purposes the curved part BD of the moment–curvature relationship is often ignored and the bending moment is assumed to increase linearly from A to C, at which time the plastic moment is reached. The ratio of the plastic moment to the yield moment is called the shape factor. The shape factor varies for different cross-sectional shapes.

Unloading from a plastic state to an elastic state is also assumed to follow the path parallel to the elastic curve. It should be realized that because of this assumption for unloading, the elastic relationship between stress and strain is no longer unique in the sense that the behavior of the material may follow any elastic curve if the material unloads from a plastic state to an elastic state. The unloading phenomenon can be demonstrated in Figure 2.4 in which the section undergoes unloading at F, from which an elastic path FG, usually parallel to AC, is assumed. When the section is fully unloaded, residual deformation corresponding to a curvature at G exists. Therefore, an elastoplastic analysis is usually performed in an incremental manner for a given history of loading in order to trace the unique states of moment and curvature in the cross section.

In reality, the exact value of M_p is difficult to obtain and its calculation is only approximate. In an experiment, a rectangular steel bar of dimensions 3 mm × 13 mm was used as a simply supported beam with a span length of 300 mm to support a centrally applied load until collapse. The beam was bending about its weaker axis. The relationship between

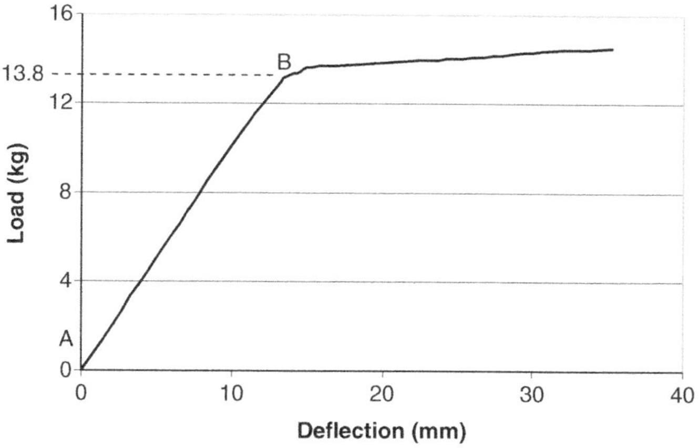

FIGURE 2.5. Load–deflection curve of a rectangular beam.

the load and the deflection at midspan is shown in Figure 2.5. This experiment shows that the load is slightly increasing at rapidly increasing deflection when the beam is loaded beyond its theoretical collapse load at B, indicating the strain-hardening effect. The theoretical collapse load is estimated to be 13.8 kg, or 135.4 N. Hence, the plastic moment of the beam section is $M_p = \frac{PL}{4} = \frac{135.4 \times 300}{4} = 10155$ Nmm. The inverse of the slope of the curve AB has been found to be 0.102 mm/N. Since the inverse of the slope of the curve AB $= \frac{L^3}{48EI}$ where $I = \frac{13 \times 3^3}{12} = 29.25\ \text{mm}^4$, hence $E = 188537\ \text{N/mm}^2$.

These mechanical properties can be used for performing elasto-plastic analysis of structures made of this type of steel bar. The mathematical expression of the general moment–curvature relationship for a rectangular section is presented next.

2.3 Moment–Curvature Relationship in an Elastic–Plastic Range

A cross section under increasing bending moment undergoes three stages of transformation in its plastification process. As shown in Figure 2.4, they are elastic (AB), elastic–plastic (BD), and fully plastic (DE).

2.3.1 Elastic Behavior

Figure 2.6 shows a small length of a beam under bending with constant curvature. The shape of the original element ABCD is transformed

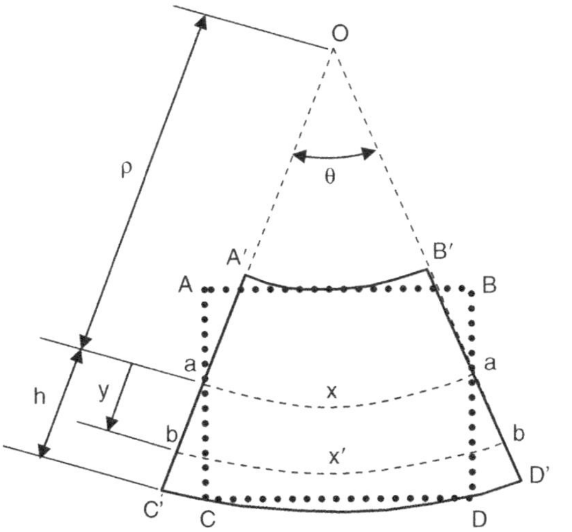

FIGURE 2.6. Part of a beam under bending.

into A′B′C′D′ as a consequence of bending so that the extreme edges A′B′ and C′D′ are both subtending a common center of circles at O. It can be seen that A′B′ contracts from AB and therefore the element fibers along A′B′ are under compression with compressive stress f_c. Similarly, the element fibers along C′D′ are under tension with tensile stress f_t. As a result, an axis exists along *aa* where the stress is zero. This is called the neutral axis at a distance ρ from the center of the circles at O.

For consistency, the sign convention is that a positive bending moment causing sagging in the element is associated with a negative radius of curvature ρ. The distance y from the neutral axis is measured positive below the neutral axis.

If the unstrained length along the neutral axis *a-a* is x and the length *b-b* at a distance y from the neutral axis is x', the axial strain can be expressed as

$$\begin{aligned} \varepsilon &= \frac{x' - x}{x} \\ &= \frac{(-\rho + y)\theta + \rho\theta}{-\rho\theta} = -\frac{y}{\rho} \end{aligned} \tag{2.1}$$

The curvature κ of a cross section is defined as

$$\kappa = \frac{1}{\rho} \tag{2.2}$$

Hence, from Equations (2.1) and (2.2), the curvature can be defined as the slope of the strain diagram shown in Figure 2.7.

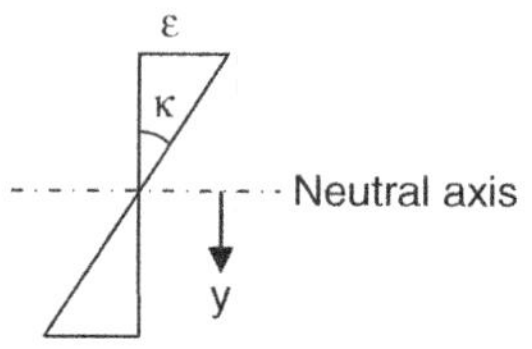

FIGURE 2.7. Curvature and strain of a cross section.

For a bending stress, f, associated with a small area ∂A in the cross section, the total compression F_c above the neutral axis is given as

$$\begin{aligned} F_c &= \int_{Top} f(\partial A) \\ &= E\kappa \int_{Top} y(\partial A) \end{aligned} \tag{2.3}$$

Likewise, the total tension F_t below the neutral axis can be derived as

$$\begin{aligned} F_t &= \int_{Bottom} f(\partial A) \\ &= E\kappa \int_{Bottom} y(\partial A) \end{aligned} \tag{2.4}$$

Note that the integrations in Equations (2.3) and (2.4) are about the neutral axis. For a cross section under pure bending, the sum of the compression and tension must vanish in order to maintain equilibrium. Hence,

$$F_c + F_t = E\kappa \int_A y(\partial A) = 0 \tag{2.5}$$

If the location of any horizontal fiber is measured as y' from a convenient axis, such as the top or bottom edge, and the location of the neutral axis is $\bar{y}$ from the same axis, Equation (2.5) can then be written as

$$E\kappa \int_A (y' - \bar{y})(\partial A) = 0 \tag{2.6a}$$

or

$$\bar{y} = \frac{\int_A y'(\partial A)}{A} \tag{2.6b}$$

where A is the total area of the cross section. Equation (2.6) is often used to locate the neutral axis numerically for simple asymmetric sections.

By taking the moment about the neutral axis, the bending moment of the whole section can be found to be

$$M = \int_A fy(\partial A) = E\kappa \int_A y^2(\partial A) \tag{2.7}$$

By defining the second moment of area (or moment of inertia) as $I = \int_A y^2(\partial A)$, the stress f at any point in the cross section can be written as

$$f = E\varepsilon = \frac{M}{I}y \tag{2.8}$$

Equation (2.8) is based on the simple bending theory in which the plane section remains plane under applied forces and has been used in elastic design method for decades. It is valid when the whole section remains elastic and the modulus of elasticity E remains constant.

2.3.2 Elastic–Plastic Behavior

To illustrate the calculation of the moment–curvature relationship beyond the elastic limit, a symmetric, rectangular section of dimensions $b \times d$ is used. When the extreme fibers of a rectangular section start to yield with $f = f_y$, the corresponding yield moment M_y, shown in Figure 2.4, is

$$M_y = \frac{f_y I}{y_{\max}} \tag{2.9}$$

or

$$M_y = f_y Z \tag{2.10}$$

where Z = elastic section modulus $= \frac{I}{y_{\max}}$ and $y_{\max}$ is the distance of the extreme fibers to the neutral axis. For a rectangular section, it can be shown that

$$Z = \frac{bd^2}{6} \tag{2.11}$$

In this case, the yield strain ε_y is related to the yield curvature κ_y by

$$\varepsilon_y = \frac{d}{2}\kappa_y \tag{2.12}$$

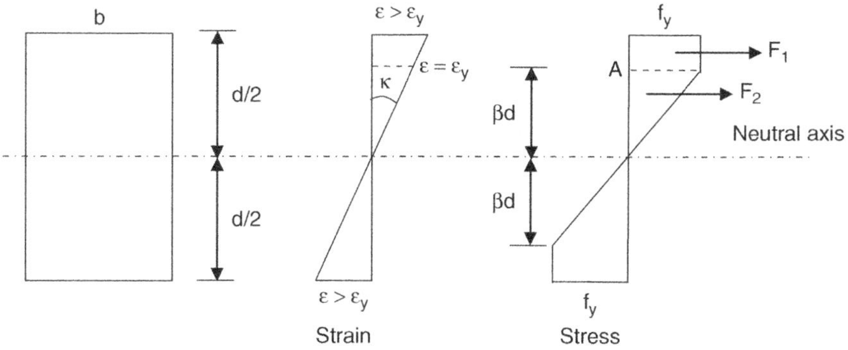

FIGURE 2.8. Elastic–plastic behavior.

A further increase in bending moment of the cross section spreads the yielding inward toward the neutral axis. Suppose that the yielding is extended to the point where the elastic core is within a distance βd from the neutral axis. The corresponding strain distribution can be calculated by the constant curvature κ shown in Figure 2.8. Hence,

$$f_y = E\kappa\beta d \tag{2.13}$$

From Figure 2.8,

$$F_1 = f_y b\left(\frac{d}{2} - \beta d\right) \tag{2.14}$$

$$F_2 = \frac{f_y}{2} b\beta d \tag{2.15}$$

Total bending moment, M, about the neutral axis is

$$\begin{aligned} M &= 2\left[F_1\left(\frac{d}{4} + \frac{\beta d}{2}\right) + F_2\left(\frac{2}{3}\beta d\right)\right] \\ &= f_y \frac{bd^2}{6}\left(\frac{3}{2} - 2\beta^2\right) \qquad (2.16) \\ &= M_y\left(\frac{3}{2} - 2\beta^2\right) \text{from Equation (2.10)} \end{aligned}$$

The moment–curvature relationship for a rectangular section has been established through Equations (2.13) and (2.16). A fully plastic section is achieved only when $\beta \rightarrow 0$. In this case, $\kappa \rightarrow \infty$ and the plastic moment for the section is $M = M_p = 1.5 M_y$. It should be noted that when the section starts to yield at $\beta = 0.5$, the yield curvature is

$$\kappa_y = 2\left(\frac{f_y}{Ed}\right) \tag{2.17}$$

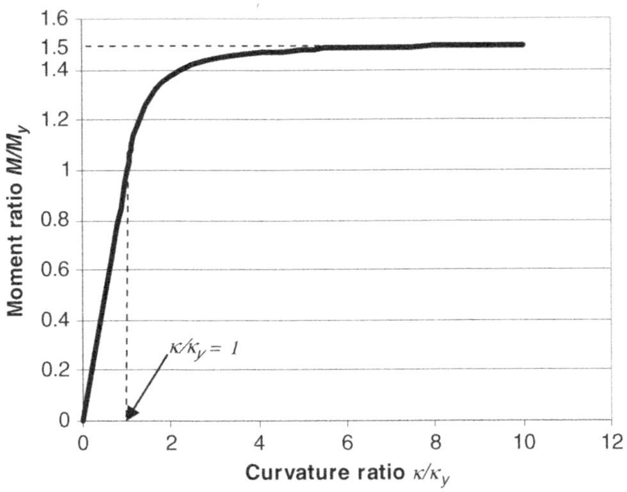

FIGURE 2.9. Dimensionless moment–curvature relationship for rectangular sections.

so that

$$\frac{\kappa}{\kappa_y} = \frac{0.5}{\beta} \text{ from Equation (2.13)} \tag{2.18}$$

Substituting Equation (2.18) into Equation (2.16), a dimensionless moment–curvature relationship in terms of the moment ratio M/M_y and curvature ratio κ/κ_y can be established. This relationship for rectangular sections is shown in Figure 2.9.

The plastic moment of any cross-sectional shape can be derived similarly in the aforementioned manner and the plastic moment can be expressed generally as

$$M_p = SM_y \tag{2.19}$$

where S = shape factor of the section. The shape factors for some common cross-sectional shapes are given in Table 2.1.

TABLE 2.1
Shape factors for common cross sections

Shape	*Shape factor*
Rectangle	1.5
Circular solid	1.7
Circular tube	1.27
Triangle	2.34
I-sections	1.1 – 1.2
Diamond	2

For some cross-sectional shapes, the derivation of the moment–curvature relationship could be tedious. With the computational tools widely available today, the moment–curvature relationship can be established easily using the numerical method. This can be done by dividing the cross section into a finite number of strips, and the moment capacity of the section is calculated by varying the value of β as in Equation (2.13) for rectangular sections. The steps for using the numerical method are described here.

1. Assign a value of β for the partial plastification of the cross section.
2. Calculate the curvature κ according to Equation (2.13).
3. Calculate strains for all strips using $\varepsilon = \kappa y$, where y is the distance of the strip from the neutral axis.
4. Calculate the bending stress of the strip using $f = E\varepsilon$.
5. Calculate the axial force in each strip using $F = A_i f$, where A_i is the area of the strip.
6. Calculate the bending moment of the strip about the neutral axis using $M = Fy$.
7. The sum of the moments from all strips is the moment capacity of the section for an assumed value of β.
8. Repeat steps 1–7, varying the value of β from 0 to the extreme fibers of the section to obtain the bending moments for varying β.

It should be noted that the sum of the axial forces is zero. This is also the condition for locating the neutral axis if the section is nonsymmetric.

The following example shows the typical procedure for using the numerical method.

Example 2.1 Determine the moment–curvature relationship and shape factor of the I section shown in Figure 2.10.

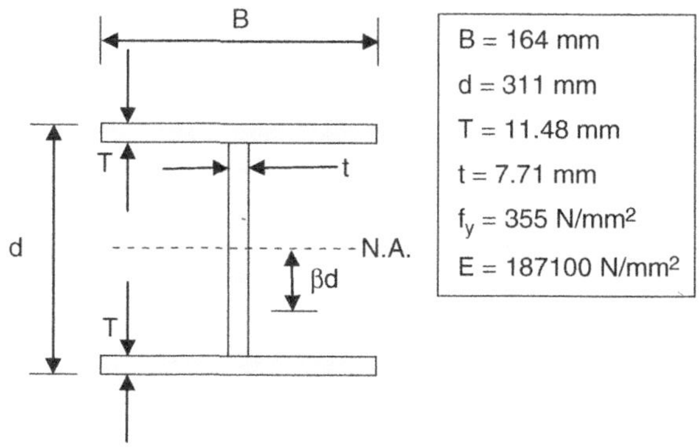

FIGURE 2.10. Shape factor of I section.

Solution. This example demonstrates the calculation of the curvatures and bending moments of an I section in tabular form, which is suitable for routine calculation using a spreadsheet such as Microsoft Excel. The section is divided into 24 horizontal strips, 4 in each of the flanges and 16 in the web. The results are shown in Table 2.2, in which only the calculations of some typical strips are shown. When the neutral axis occurs at a point within a strip, the strip is further divided into two strips, one below and one above the neutral axis, and the calculation is performed for each strip accordingly. The results of the bending moments for varying elastic core distance βd are shown in Table 2.3. The resulting shape factor for this section is 1.128. It should be noted that for nonsymmetric sections, the location of the neutral axis can be found such that the sum of the axial forces of the strips in Table 2.2 is zero.

TABLE 2.2
Moment–curvature computation for $\beta d = 150$ mm from neutral axis

Strip No.	*Strain at centroid of strip*	*Bending stress (MPa)*	*Axial force in strip due to bending (kN)*	*Moment about neutral axis (kNm)*
1	−0.00195	−355.0	−167.1	25.7
..	..	..	..	..
..	..	..	..	..
21	0.00184	344.2	162.0	23.6
22	0.00188	351.0	165.2	24.5
23	0.00191	355.0	167.1	25.3
24	0.00195	355.0	167.1	25.7
		Total = 0		234.4

TABLE 2.3
Results of bending moments for varying elastic core distance βd

Elastic core βd (mm)	*κ (mm^{-1})*	*Moment about neutral axis (kNm)*
0	∞	257
15	1.2649×10^{-4}	256.6
30	6.3246×10^{-5}	256.1
60	3.1623×10^{-5}	253.6
90	2.1082×10^{-5}	249.4
120	1.5812×10^{-5}	243.8
150	1.2649×10^{-5}	234.4
155.5	1.2202×10^{-5}	227.8

Distance of neutral axis from bottom edge = 155.5 mm

$\kappa = 355/(187100 \times 150) = 1.2649 \times 10^{-5}$.

The moment–curvature relationship can be used to measure the extent of the inelastic zone in a member. Take the cantilever beam of length L subjected to a point load P at its free end shown in Figure 2.11 as an example. The location of C where the extreme fibers of the beam start to yield corresponds to the section having a yield bending moment M_y.

The extent of the elastic region is given as

$$x_e = \frac{M_y}{P} \tag{2.20}$$

Equation (2.20) is valid so long as $PL \leq M_P$. When the beam is at imminent collapse caused by a collapse load $P = P_c$, the section at B is fully plastic for which the length of the elastic region is a minimum, denoted by x_{emin} and calculated from Equation (2.20). That is,

$$P_c L = M_P \tag{2.21}$$

By using $P = P_c$ and $x_e = x_{emin}$ in Equations (2.20) and (2.21),

$$\frac{x_{emin}}{L} = \frac{M_y}{M_P} = \frac{1}{S} \tag{2.22}$$

Equation (2.22) shows that, for this example, the minimum length of the elastic zone, or the maximum extent of the inelastic zone, is inversely proportional to the shape factor of the section. In most structural shapes used in structures, this is generally true. In general, the deflection of the structural member depends on the extent of the inelastic zone. For sections with low values of shape factor, such as I sections, the difference in the member's deflections calculated with and without an inelastic zone is very small. In plastic analysis of general structures, it would be a highly complex task if the inelastic zone such as that shown in Figure 2.11 is considered. Naturally, a simplified and

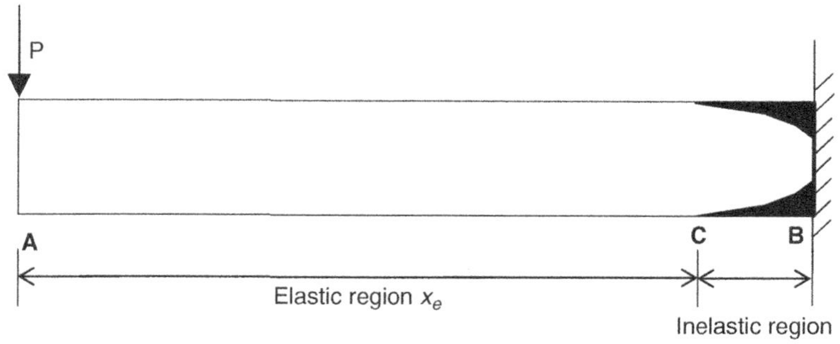

FIGURE 2.11. Spread of inelastic zone.

easy-to-use material constitutive relationship that bears some resemblance to the actual stress–strain curve is desired. For this reason, the inelastic zone is usually ignored in engineering practice and the section is assumed to be elastic until the plastic moment is reached. This is equivalent to following the bilinear moment–curvature relationship ACE shown in Figure 2.4. When the section reaches its plastic moment under this idealization, the section becomes a plastic hinge occupying infinitesimal length for the inelastic zone. This forms the basis of the elastoplastic theory for elastic A-perfectly plastic material using the plastic hinge concept, which is adopted for the subsequent development of the work described in this book.

Example 2.2 Determine the inelastic zone length x_p of the simply supported beam shown in Figure 2.12 when the section under the load becomes fully plastic. The beam has a uniform rectangular section with a plastic moment capacity of 165 kNm.

Solution. The section starts to yield at a bending moment of $M_y = \dfrac{M_p}{1.5} = 110$ kNm. The bending moment diagram of the beam is shown in Figure 2.13. If the lengths of the inelastic zone are x_{p1} to the left and x_{p2} to the right of the applied load, then from the geometry of the bending moment diagram,

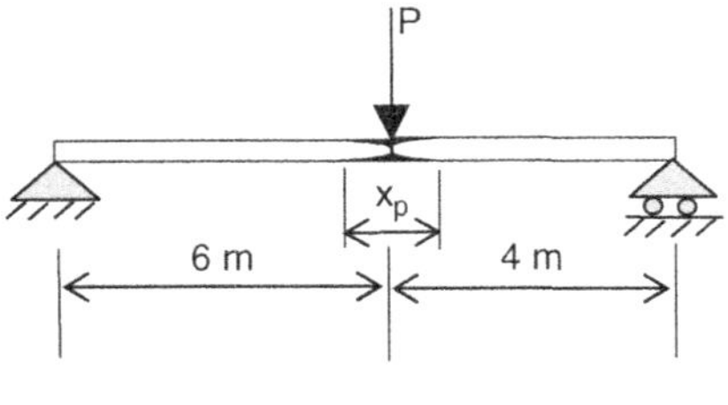

FIGURE 2.12. Inelastic zone of simply supported beam.

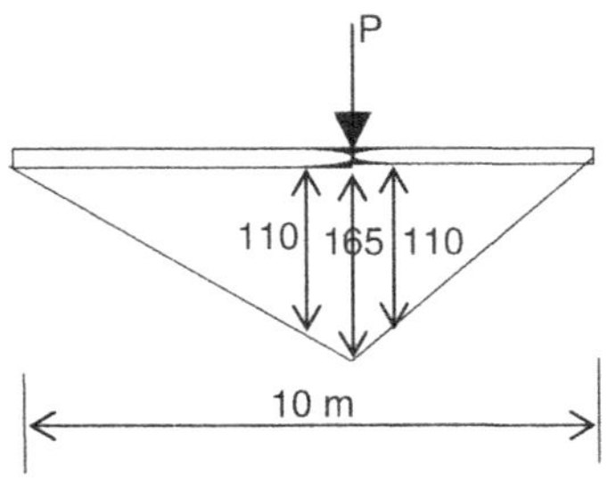

FIGURE 2.13. Bending moment diagram.

$$\frac{6-x_{p1}}{6}=\frac{110}{165} \qquad \therefore x_{p1}=2 \text{ m}$$

$$\frac{4-x_{p2}}{4}=\frac{110}{165} \qquad \therefore x_{p2}=1.33 \text{ m}$$

Total length of the inelastic zone is 3.33 m.

2.3.3 Fully Plastic Section

Suppose that the stress distribution of a cross section is as shown in Figure 2.14 so that the total tensile force above the neutral axis X–X is T and the total compressive force below the neutral axis is C. For a fully plastic section of arbitrary shape under pure bending, equilibrium requires that the net axial force be zero. Hence,

$$T=\int_{A'} f_y \partial A = C \tag{2.23}$$

where A' is the area above or below the neutral axis.

Equation (2.23) is used to find the location of the neutral axis. Assume that the yield stress f_y is the same throughout the section, then Equation (2.23) becomes

$$A_t=\int_{A_t} \partial A = A_c = \int_{A_c} \partial A \tag{2.24}$$

where A_t and A_c are areas above and below the neutral axis, respectively. Therefore, for sections made of materials with uniform yield stress, the neutral axis X–X is also called the equal area axis, which divides the cross section into two equal parts. For symmetric sections, the equal area axis coincides with the centroid of the section about which the plastic section modulus is calculated. For nonsymmetric sections, the locations of the equal area axis and the centroid of the section are different.

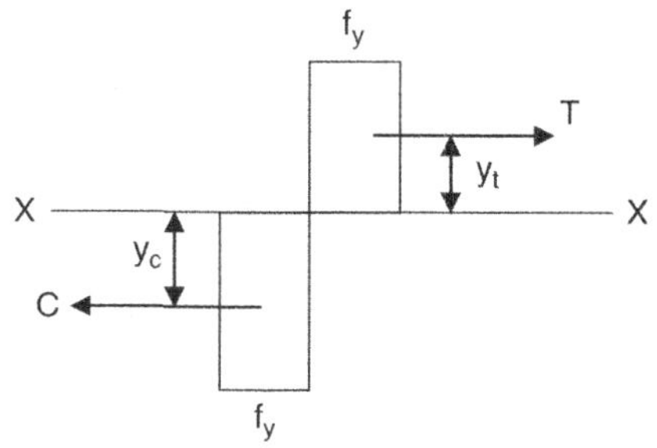

FIGURE 2.14. Stress distribution for a fully plastic section.

By taking the moment about X–X,

$$M_p = Ty_t + Cy_c = (A_t y_t + A_c y_c) f_y = Z_s f_y \tag{2.25}$$

where Z_s = the first moment of area about the equal area axis = plastic section modulus.

For the rectangular section shown in Figure 2.15, the plastic section modulus is given by

$$Z_s = 2 \times \left(\frac{bd}{2} \times \frac{d}{4}\right) = \frac{bd^2}{4} \tag{2.26}$$

Comparing Equation (2.26) with the elastic section modulus Z and from Equation (2.19),

$$S = \frac{M_P}{M_y} = \frac{Z_s f_y}{Z f_y} = 1.5 \tag{2.27}$$

which is the same as that obtained from Equation (2.16) for $\beta \rightarrow 0$.

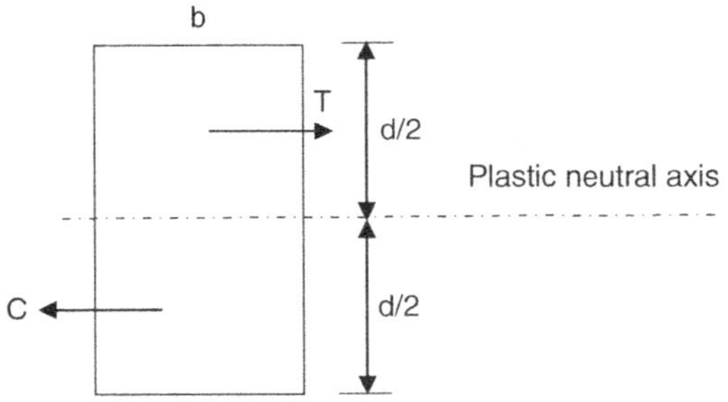

FIGURE 2.15. Plastic section modulus of a rectangular section.

Example 2.3 Determine the plastic section modulus Z_s for the I section shown in Figure 2.16.

Solution

$$Z_s = 2[(200 \times 40 \times 220) + (200 \times 40 \times 100)]$$
$$= 5120 \times 10^3 \text{ mm}^3$$

Example 2.4 Determine the plastic moment M_p and plastic section modulus Z_s for the T section shown in Figure 2.17 if (a) $f_y = 320$ MPa for the whole section and (b) $f_y = 320$ MPa for the top flange and $f_y = 250$ MPa for the web.

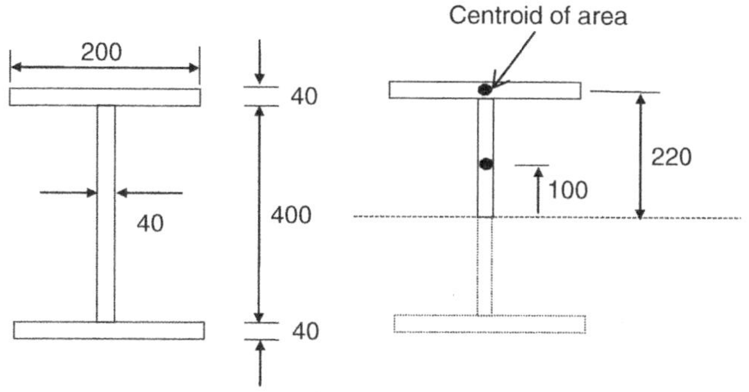

FIGURE 2.16. Plastic section modulus of an I section.

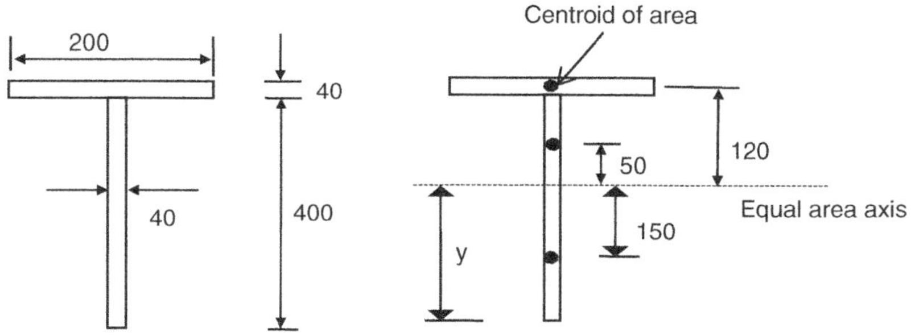

FIGURE 2.17. Plastic section modulus of a T section.

Solution

(a) Total area $= (200 \times 40) + (400 \times 40) = 24000\ \text{mm}^2$

$$(40 \times y) = 24000/2$$
$$\therefore y = 300\ \text{mm}$$
$$Z_s = (200 \times 40 \times 120) + (100 \times 40 \times 50) + (300 \times 40 \times 150)$$
$$= 2960 \times 10^3\ \text{mm}^3$$
$$M_p = 320 \times 2960 \times 10^3 = 947.2\ \text{kNm}$$

(b) Total force $= (200 \times 40) \times 320 + (400 \times 40) \times 250 = 6560\ \text{kN}$

For half of the force,

$$(40 \times y) \times 250 = 6560000/2$$
$$y = 328\ \text{mm}$$
$$M_p = 320 \times (200 \times 40 \times 92) + 250 \times (72 \times 40 \times 36)$$
$$+ 250 \times (328 \times 40 \times 164)$$
$$= 799\ \text{kNm}$$

2.4 Plastic Hinge

According to the bilinear moment–curvature idealization shown in Figure 2.4, a section attaining its plastic moment capacity undergoes plastic rotation without any further increase in bending moment. In other words, the section behaves like a real hinge while possessing a fully plastic moment. This hinge behavior, typically pertaining to a plastic hinge, enables a structure to be analyzed continuously by inserting a plastic hinge at any section reaching its plastic moment. This is the basic concept for elastoplastic analysis to be performed on structures using the hinge-by-hinge concept. In tracing the formation of the plastic hinges, the structure becomes increasingly flexible until its stiffness is reduced to such a small value that imminent collapse occurs.

For an indeterminate structure under increasing loading, the magnitude of the increase in loads can be calculated by considering the attainment of plastic moments in sections in an elastoplastic analysis. Take a fixed-end beam under a point load shown in Figure 2.18 as an example. The collapse mechanism of the beam requires the formation of three plastic hinges at A, B, and C.

In carrying out the elastoplastic analysis for this beam, or for any structure in general, the stiffness-deteriorating nature of the structure can be visualized by plotting the variation of the load with deflection at a point. For the fixed-end beam, the variation of the load P with the vertical deflection at B is plotted and shown in Figure 2.19.

In plotting the load–deflection curve shown in Figure 2.19, it should be noted that

- each "black dot" represents a plastic hinge at a section in a fully plastic state; the plastic hinge has attained a bending moment equal to its plastic moment;
- the elastic state of a structure corresponds to a load level below the first plastic hinge. Analysis below this load level is called *elastic analysis*;
- the elastic–plastic state of a structure corresponds to any load level between the first and the last plastic hinges. Analysis at this load level is called *elastoplastic analysis*;

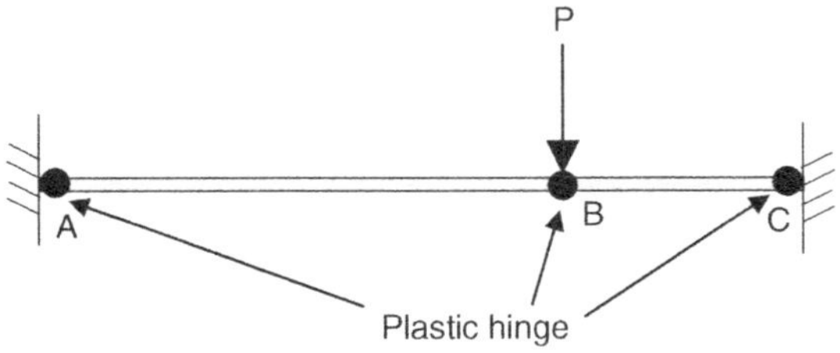

FIGURE 2.18. Collapse mechanism with plastic hinges of a fixed-end beam.

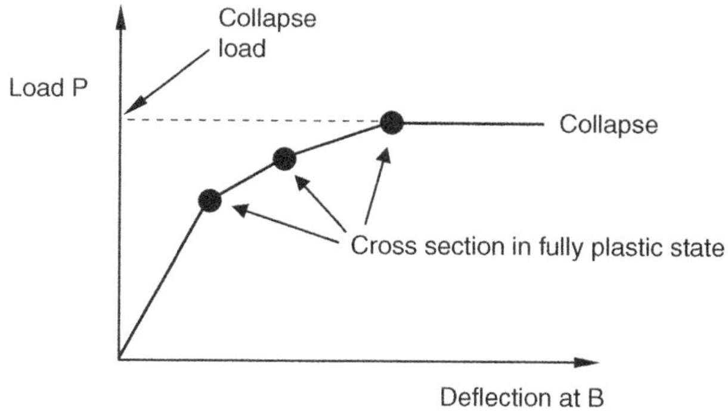

FIGURE 2.19. Load–deflection curve of a fixed-end beam.

- the behavior of a structure between the formation of consecutive plastic hinges is elastic and can be analyzed elastically;
- a fully plastic state of a structure corresponds to a load level at which the structure collapses. At this load level, analysis stops;
- the slope of the curve indicates the relative stiffness of the structure; the stiffness decreases as more sections become plastic hinges. At collapse, the stiffness of the structure is zero.

2.5 Plastic Design Concept

Plastic design makes use of the reserve strength beyond the elastic state of the structure. The structure's reserve strength, which allows structural members to be loaded without failure when their maximum bending capacity is reached, is utilized through the elastic–plastic state when the loading is increasing. As a result, a more economical design due to material saving can be achieved when using the plastic design method. Plastic design can be viewed as a means whereby the ability of moment redistribution of steel structures is utilized when the structures are loaded beyond their elastic state. It may be noted that in adopting a bilinear moment–curvature relationship for steel, the beneficial effect of strain hardening of the material is ignored. Thus, as far as plastic analysis is concerned, the theoretical plastic collapse load is always less than the true load and the resulting design is always slightly conservative.

2.6 Comparison of Linear Elastic and Plastic Designs

For elastic design, each of the members in the structure must have a design bending moment capacity (ϕM_s) greater than the design moment (M^*) obtained from an elastic analysis. The bending moment capacity, calculated by including the appropriate capacity factor ϕ and

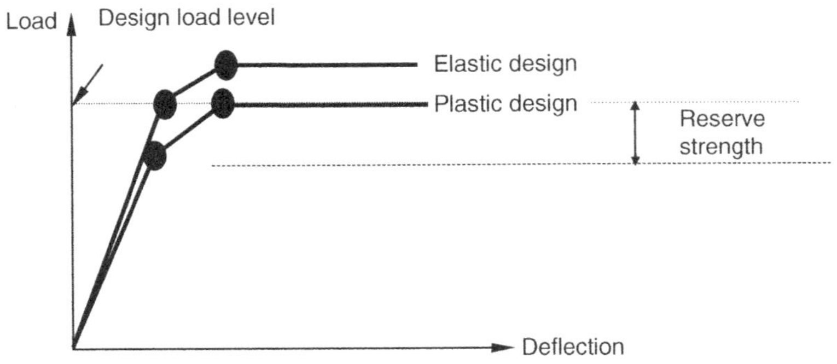

FIGURE 2.20. Comparison of elastic design and plastic design.

other design considerations, is often used to represent the plastic moment M_p in plastic analysis. Under the design loading, if one of the members meets the design requirement such that $\phi M_s = M^*$, the first plastic hinge occurs exactly at the design load level along the elastic design curve shown in Figure 2.20. In most cases, the choice of the section for elastic design leads to $\phi M_s > M^*$ so that the first plastic hinge of the structure always occurs at a load level above the design loading.

In contrast, plastic design requires that the last plastic hinge occurs at or above the design load level. It is clear from Figure 2.20 that if both elastic and plastic designs satisfy the same design loading, the plastic design method requires a lighter structure with smaller member size by utilizing the reserve strength of the structure. It is noted that for a structure with a high degree of statical indeterminacy, the reserve strength is large. Therefore, the benefit of using plastic design is greater for structures with high degrees of statical indeterminacy. However, for determinate structures that require only one plastic hinge to induce a collapse, there is no difference between elastic and plastic design methods.

2.7 Limit States Design

Limit states design (LSD), also termed load and resistance factor design (LRFD) in the United States, is based on realistic loading conditions and material properties as opposed to allowable stress design (ASD), which is mainly based on prescribed loading and stress limits. Although the linear elastic design method can be applied to both LSD and ASD, only the latter is truly elastic. While linear elastic analysis is performed using the linear load–displacement relationship, elastic bending theory is applied to ASD for stress calculation, whereas for LSD, plastic bending theory is used for moment capacity calculation of the section.

For steel structures, two major limit states need to be considered for general design: the ultimate limit state and the serviceability limit state. There are other limit states that may need special treatment and are usually classified under "accidental loadings" in design codes. In the present context, only ultimate and serviceability limit states are dealt with. Analysis and design for structures at elevated temperatures are also given, albeit briefly.

Ultimate limit state design requires that Equation (2.28) be satisfied:

$$\sum_i \gamma_i P_i = S_n^* \leq \phi R_n \qquad (2.28)$$

in which P_i = nominal load of type I, including dead, live, wind, and snow; γ_i = corresponding load factor; S_n^* = member actions including axial force, moment, and shear for member n; ϕ = capacity factor; R_n = nominal member capacity. For plastic design, the loads are increased proportionally by a common factor α_c such that

$$\alpha_c \sum_i \gamma_i P_i = S_n^* \leq \phi R_n \text{ and } \alpha_c \geq 1.0 \quad \text{at collapse of structure} \qquad (2.29)$$

Equation (2.28) implies that an analysis is performed on the structure that is subjected to the factored loads, whereas Equation (2.29) implies that the analysis is performed by increasing the common load factor α until the structure collapses at $\alpha = \alpha_c$. In both Equations (2.28) and (2.29), the member actions S_n^* may include both the material and the geometric effect. In plastic analysis as described in this book, the geometric effect is not included and is treated separately in design.

2.8 Overview of Design Codes for Plastic Design

Most countries use similar rules for plastic design. In essence, specifications in plastic design codes require the construction materials to have adequate ductility for the plastic moment to be fully developed and sustained until collapse. In addition, for steel material the plastic hinge should be able to undergo plastic rotation without suffering from local buckling. Details of the design rules are discussed in Chapter 8. Plastic design codes of practice being used in Australia, the United States, and Europe are described as they represent the current design practice in many other countries.

Plastic design rules developed in the 1960s and 1970s have been used for a long time. These rules were written on the basis of simple plastic theory, which was the main concept for analysis at the time. The simple plastic theory can be applied only to simple structures using mainly manual methods for analysis. For instance, the 'European recommendations for steel construction' state that "...clauses will generally apply to: -simple or continuous beams; -one and two storey frames, braced or unbraced." Most of these rules for plastic design are still stipulated in current design codes, although the advent of computers has allowed advanced nonlinear analysis, including second-order effects, to be admitted for design.

In the United States, both LRFD and ASD design methods are used by engineers. Both the LRFD specification (AISC, 1999)[12] and the ASD specification (AISC, 1989)[13] have been superseded by the combined LRFD–ASD national standard (AISC, 2005).[14] The latest specification for structural steel buildings includes an appendix for 'Inelastic analysis and design,' which is essentially a collection of clauses for plastic design rules published in the earlier specification.

In Australia, specifications for steel structures design are published in AS4100.[15] Clauses for plastic design are intertwined with those for elastic design. Second-order effects on bending moment due to geometry change of both the structural members and the structure are taken into account by using a moment amplification factor. This is, in effect, a reduction factor for the collapse load in plastic analysis.

In Europe, a set of unified structural codes has been published by a concerted effort through the European community. Eurocode 3 (CEN, 2005)[16] is the specification for steel structures design. Similar to AS4100, the rules for plastic design are intertwined with those for elastic design. Different types of analysis for plastic design, including rigid-plastic, elastic–plastic, and nonlinear plastic, are allowed using plastic hinges or inelastic zone models.

2.9 Limitations of Plastic Design Method

Although the plastic design method provides some advantages over the elastic design method, there are limitations to its use. Some of these limitations are described in this section.

For plastic design, all cross sections must be able to sustain the plastic moment M_p without showing any sign of local buckling. To achieve this, the sections must be compact (or class 1 and class 2 cross sections in Eurocode 3). A typical plastic hinge attaining its plastic moment without local buckling effect is shown in Figure 2.21a,

FIGURE 2.21. Beam under bending with and without local buckling effects.

whereas a beam subject to local buckling under bending is shown in Figure 2.21b.

Unless using an advanced nonlinear plastic analysis for design, the effect of lateral–torsional buckling on plastic behavior of the structure is not considered. Therefore, all members designed by the plastic method should be provided with adequate lateral restraints to prevent lateral buckling.

The steel material must be ductile enough to undergo plastic rotation without failure. This ductility and other material requirements are provided in most design codes. For example, the following rules stipulated in AS4100 must be satisfied:

1. Hot-formed, doubly symmetric I sections are used with $f_y \leq 450$ MPa.
2. The stress–strain relationship for the steel material must have the characteristics shown in Figure 2.22.
3. No fatigue assessment is required.

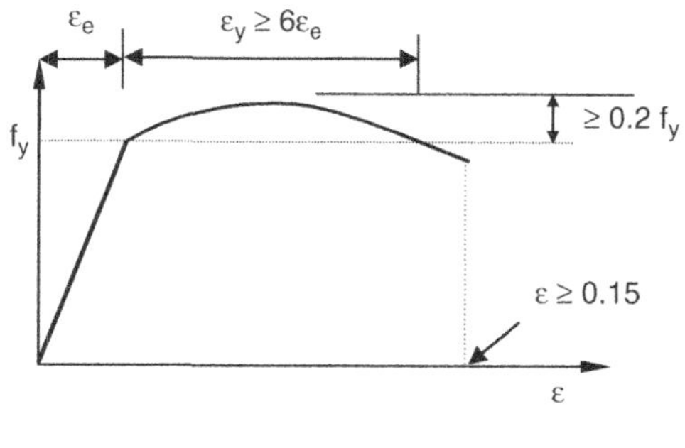

FIGURE 2.22. Ductility requirements for plastic design.

Problems

2.1. Determine the moment–curvature relationship of the rectangular section shown in Figure P2.1.

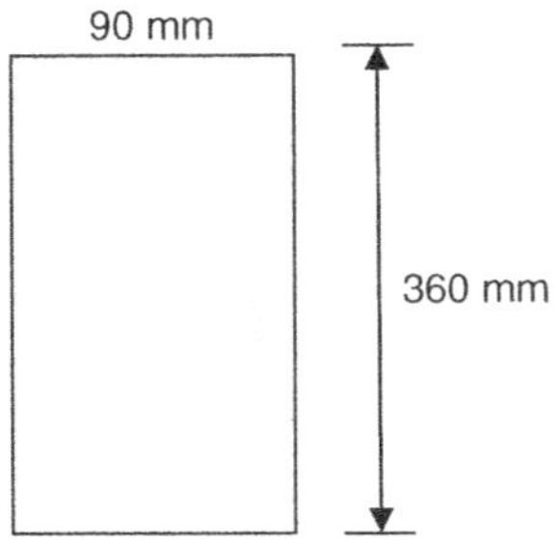

FIGURE P2.1. Problem 2.1.

2.2. Determine the inelastic zone length at midspan for the beam subject to a uniform distributed load shown in Figure P2.2. The plastic moment of the rectangular section is 60 kNm. Assume a linear moment-curvature relationship up to plastic moment.

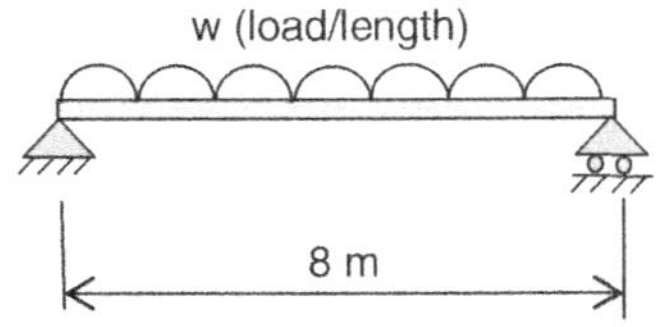

FIGURE P2.2. Problem 2.2.

2.3. Determine the inelastic zone lengths at A and B for the beam of uniform rectangular section shown in Figure P2.3 when (a) the section at A becomes fully plastic and (b) when both sections at A and B become fully plastic. The plastic moment for the beam = 120 kNm. Elastic moment at A = $\frac{3PL}{16}$ and at B = $\frac{5PL}{32}$. L = length of beam. Assume a linear moment-curvature relationship up to plastic moment.

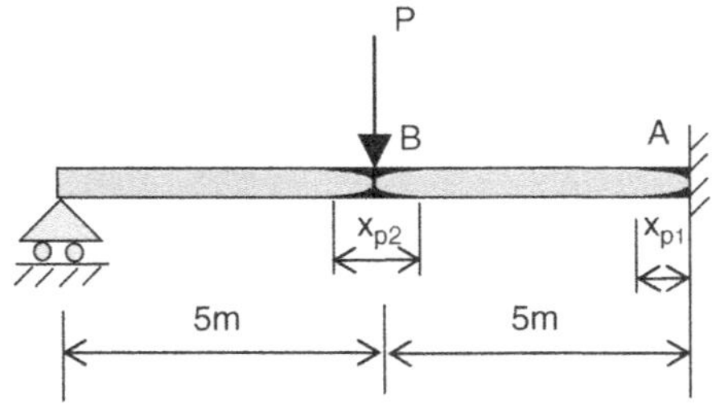

FIGURE P2.3. Problem 2.3.

2.4. Determine the inelastic zone lengths at A, B, and C for the fixed-end beam of the uniform I section shown in Figure P2.4 when (a) the section at A becomes fully plastic and (b) when sections at A, B, and C all become fully plastic. Plastic moment for the beam = 360 kNm. Shape factor S = 1.2. Assume a linear moment-curvature relationship up to plastic moment.

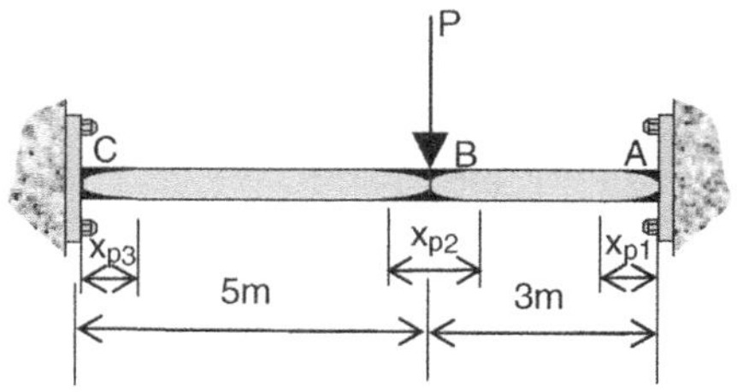

FIGURE P2.4. Problem 2.4.

2.5. Determine the location of the plastic neutral axis y and the plastic moment for the section shown in Figure P2.5 if (a) the section has a uniform yield stress of 300 MPa and (b) the top and bottom flanges have a yield stress of 300 MPa and the web has a yield stress of 200 MPa.

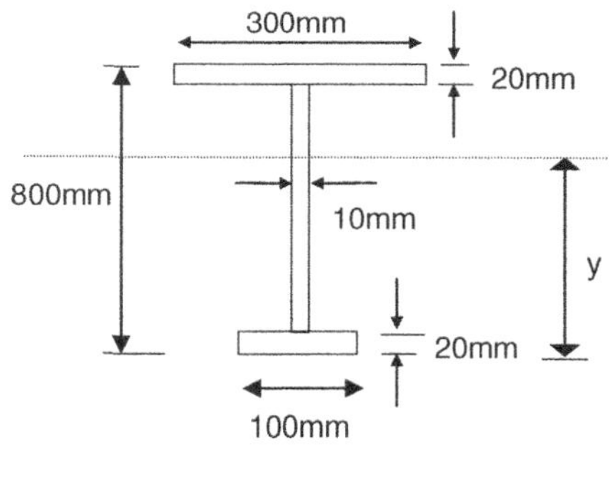

FIGURE P2.5. Problem 2.5.

2.6. Determine the location of the plastic neutral axis and the plastic section modulus for the section of uniform yield stress shown in Figure P2.6. a = 140 mm, b = 410 mm, c = 360 mm.

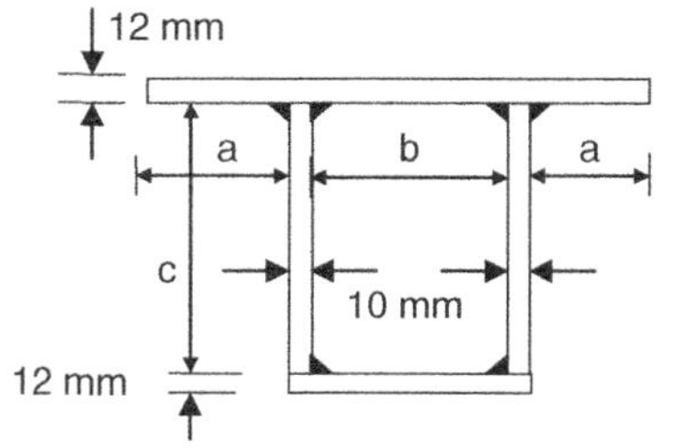

FIGURE P2.6. Problem 2.6.

Bibliography

1. Marcal, P. V., and King, I. P. (1967). Elastic-plastic analysis of two-dimensional stress systems by the finite element method. *Int. J. Mech. Sci.*, **9**, pp. 143–155.
2. Yamada, Y., and Yoshimura, N. (1968). Plastic stress-strain matrix and its application for the solution of elastic-plastic problems by the finite element method. *Int. J. Mech. Sci.*, **10**, pp. 343–354.
3. Zienkiewicz, O. C., Valliappan, S., and King, I. P. (1969). Elastoplastic solution of engineering problems, 'Initial stress' finite element approach. *Int. J. Num. Meth. Eng.*, **1**, pp. 75–100.
4. Ueda, Y., Yamakawa, T., Akamatsu, T., and Matsuishi, M. (1969). A new theory on elastic-plastic analysis of framed structures. *Tech. Repts.*, Osaka Univ., **19**, pp. 263–275.
5. Nigam, N. C. (1970). Yielding in framed structures under dynamic loads. *J. Eng. Mech. Div.*, ASCE, **96**, pp. 687–709.
6. Livesley, R. K. (1964). *Matrix methods of structural analysis*, London, England. Pergamon Press.
7. Davies, J. M. (1966). The response of plane frameworks to static and variable repeated loading in the elastic-plastic range. *St. Eng.*, **44**(8), pp. 277–283.
8. Majid, K. I. (1972). *Non-linear structures*, London, UK. Butterworth & Co. Ltd.
9. Franchi, A., and Cohn, M. Z. (1980). Computer analysis of elastic-plastic structures. *Comp. Meth. Appl. Mech. Eng.*, **21**, pp. 271–294.
10. Maier, G., and Munro, J. (1982). Mathematical programming applications to engineering plasticity analysis. *Appl. Mech. Reviews*, **35**, pp. 1631–1643.
11. Tin-Loi, F., and Pang, J. (1993). Elastoplastic analysis of structures with nonlinear hardening: A nonlinear complementarity approach. *Comp. Meth. Appl. Mech. Eng.*, **107**, pp. 299–312.
12. AISC (1999). *Load and resistance factor design specification for structural steel buildings*. Chicago: American Institute of Steel Construction.
13. AISC (1989). *Specification for structural steel buildings—allowable stress design and plastic design*. Chicago: American Institute of Steel Construction.
14. ANSI/AISC 360-05. (2005). *Specification for structural steel buildings*. Chicago, USA American Institute of Steel Construction.
15. Standards Australia. (1998). *AS4100 Steel structures*. Standards Australia.
16. CEN (2005). *EN 1993-1-1. Design of steel structures - Part 1-2*. CEN.

CHAPTER 3

Plastic Flow Rule and Elastoplastic Analysis

3.1 General Elastoplastic Analysis of Structures

Chapter 2 shows that a plastic hinge sustaining the full plastic moment behaves like a real hinge while undergoing plastic rotation. This implies that the behavior of a steel structure between the formations of consecutive plastic hinges is elastic and can be simulated using the linear elastic analysis method as described in Chapter 1. Take a structure with plastic hinges formed at load levels 1, 2, and 3 shown in Figure 3.1 as an example.

Recall that the equilibrium of a structure is expressed as

$$\{F\} = [K]\{D\} \tag{3.1}$$

For elastoplastic analysis, a common load factor is assigned to the load vector $\{F\}$ such that by increasing this common load factor, the formation of the plastic hinges at 1, 2, 3, etc., shown in Figure 3.1 can be traced incrementally until the structure collapses. If the common load factor corresponding to the formation of the first plastic hinge at 1 (at a section in one of the members of the structure) is α_1, the solution of Equation (3.1) for the displacement increment vector $\{\Delta D\}_1$ can be written as

$$\{\Delta D\}_1 = [K]^{-1}\alpha_1\{F\} \tag{3.2}$$

The incremental member force vector at load level 1 can be calculated using

$$\{\Delta P\} = [K_e][T]^t\{\Delta D_e^g\} \tag{3.3}$$

where the local displacement increment vector $\{\Delta D_e^g\}$ is extracted from $\{\Delta D\}_1$.

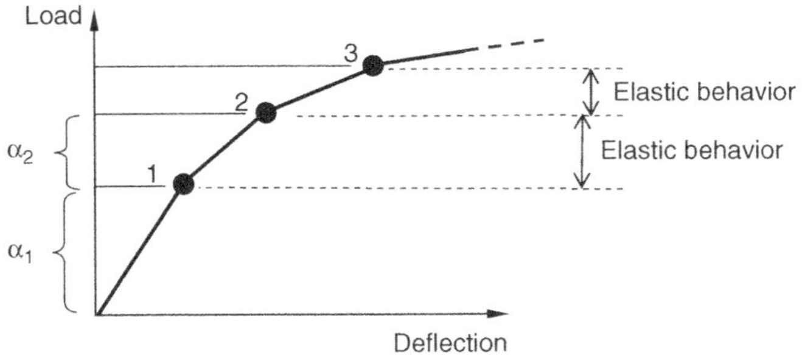

FIGURE 3.1. Hinge formation of a structure.

Subsequent analysis simulating the behavior of the structure subject to increment of loads between load levels 1 and 2 requires modification to the stiffness of the member at the point where the formation of the plastic hinge occurs. This is done by treating the plastic hinge as an internal pin as described in Section 1.12 of Chapter 1, in which two methods of pin modeling have been presented. These methods of pin modeling can also be applied in an incremental elastoplastic analysis.

After modification to the member stiffness matrix where the hinge modeling is made, the incremental load factor can be found between load levels 1 and 2, hereafter called α_2 for which the next plastic hinge at level 2 occurs. The solution for α_2 is obtained by solving Equation (3.4):

$$\{\Delta D\}_2 = [K_P]^{-1}\alpha_2\{F\} \tag{3.4}$$

where $[K_P]$ is the modified structure stiffness matrix, taking into account the formation of the plastic hinge at 1.

The aforementioned procedure can be repeated until collapse occurs when the determinant of the structure stiffness matrix becomes zero. Provided that the same load vector $\{F\}$ is used for each incremental step, the collapse load factor of the structure can be obtained as the sum of the incremental ones and is given as

$$\alpha_{col} = \alpha_1 + \alpha_2 + \ldots\ldots. \tag{3.5}$$

It can be seen that an important aspect in elastoplastic analysis is determination of the load level at which the plastic hinge occurs. For a member with yielding only under pure bending, the criterion for formation of the plastic hinge is straightforward and is based on that described in Chapter 2. However, apart from bending moments, structural members in most structures are also subjected to different types of actions such as axial forces, shear forces, and torsion. These forces may affect and usually reduce the plastic moment capacity of the section in some or all members and subsequently affect the plastic collapse loads or load factor of the structure. The influence of these forces on the plastic moment and the corresponding stiffness of the structural elements are described in this chapter. The derivation leads to use of the concept of yield surface whereby the relationship between the element stiffness matrix and force interactions can be established. Details of the procedures to deal with the reduced plastic moment in relation to the calculation of the collapse load factor α_{col} in incremental elastoplastic analysis are given in Chapter 4.

3.2 Reduced Plastic Moment Capacity Due to Force Interaction

The presence of axial and shear forces in a cross section reduces its plastic moment capacity. In many cases, the effect of axial force is significant, even in moderately complex structures, whereas the effect of shear force is usually small and can be ignored. The interaction formula for the reduced plastic moment capacity of a rectangular section due to axial force is derived. For the commonly used I sections, the interaction formula is given without derivation. Use of the interaction formula leads to the important concept of yield surface in plasticity. Torsion may also affect the plastic moment capacity of a section. However, this is the case mainly for slender three-dimensional structures, which are not dealt with here.

3.2.1 Axial Force

The coexistence of axial force and bending moment occurs when, for example, an axial force is acting eccentrically about the centroid of the cross section. This is equivalent to the simultaneous application of both the axial force acting through the centroid and the eccentric moment bending about the elastic neutral axis of the section. In this case, plastification starts at the extreme point on one side of the elastic neutral axis and spreads toward the final plastic neutral axis as the axial force increases. The formula for the moment–axial force relationship depends on the shape of the cross section. The following shows the derivation of the reduced plastic moment M_{pr} due to axial force for a rectangular cross section. The derivation holds when the same yield stress applies to both compression and tension.

Rectangular Sections

When a compressive axial force N is applied to a rectangular section subject to bending, its effect is to increase the compression zone of the section, shifting the plastic neutral axis to a location as shown

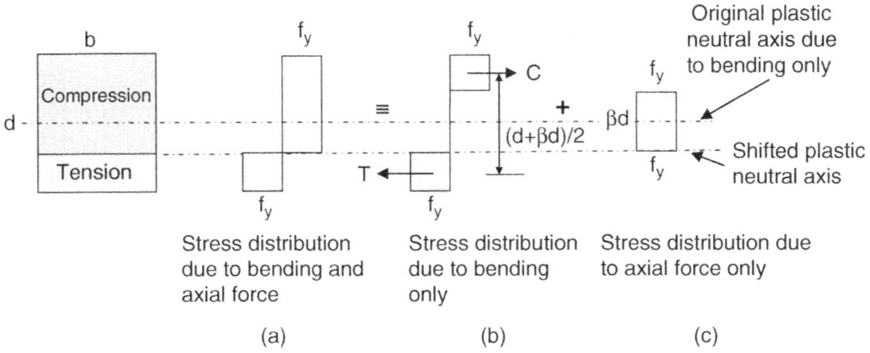

FIGURE 3.2. Effect of axial force on a rectangular section.

in Figure 3.2. The final stress distribution, shown in Figure 3.2a, can be split into two parts, one in pure bending as in Figure 3.2b and the other in pure axial as in Figure 3.2c.

From Figure 3.2,

$$N = (b\beta d)f_y \quad \text{or} \quad \beta = \frac{N}{bdf_y} = \frac{N}{N_p} \tag{3.6}$$

where $\beta \leq 1$ and N_p = squash load of the section. The value of β can be considered as the magnitude of the applied axial force N in the section relative to the squash load. The total compressive force C, equal to the total tensile force T, due to pure bending, is given by

$$C = T = bd\frac{(1-\beta)}{2}f_y \tag{3.7}$$

Hence, the reduced plastic moment capacity, M_{pr}, is

$$\begin{aligned} M_{pr} &= C \times \frac{(d+\beta d)}{2} = \frac{bd^2}{4}f_y(1-\beta^2) \\ &= M_p(1-\beta^2) \end{aligned} \tag{3.8}$$

where M_p is the plastic moment capacity of the section under pure bending.

I Sections

The derivation of M_{pr} for I sections follows a similar procedure as for the rectangular sections given earlier. There are two cases to be considered: the neutral axis in the web and the neutral axis in the flange.

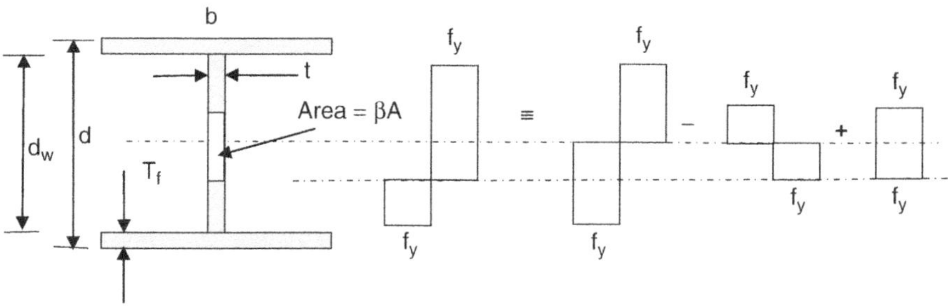

FIGURE 3.3. Effect of axial force on an I section.

(i) Neutral axis in the web. This is equivalent to a section carrying an axial force acting on an area equal to βA where $\beta = \dfrac{N}{N_p} \le \dfrac{t(d-2T_f)}{A}$ and $A = 2bT_f + t(d-2T_f)$. The section and the stress distribution are shown in Figure 3.3.

It can be derived that

$$M_{pr} = M_p\left(1 - \frac{\beta^2 A^2}{4tZ_S}\right) \tag{3.9}$$

where Z_S is the plastic section modulus given by

$$Z_S = bT_f(d - T_f) + t\left(\frac{d}{2} - T_f\right)^2 \tag{3.10}$$

(ii) Neutral axis in the flange. In this case, the axial force is acting on an area equal to βA where $\beta = \dfrac{N}{N_p} > \dfrac{t(d-2T_f)}{A}$. The corresponding reduced plastic moment capacity can be derived as

$$M_{pr} = M_p\left\{1 - \frac{1}{4Z_S}\left[td_w^2 + \frac{(\beta A - td_w)^2}{b} + 2(\beta A - td_w)d_w\right]\right\} \tag{3.11}$$

3.2.2 Shear Force

The magnitude of shear force in a structure is usually so small that it has little effect on the plastic moment capacity. In the exceptional case of high shear force, the interaction equation can be derived from the von Mises yield criterion. For a section with bending stress σ and shear stress τ, the von Mises yield criterion gives

$$\sigma^2 + 3\tau^2 = f_y \tag{3.12}$$

For pure shear for which $\sigma = 0$, the shear yield stress τ_y is, according to Equation (3.12),

$$\tau_y = \frac{f_y}{\sqrt{3}} \tag{3.13}$$

For a rectangular section, a lower bound solution, details of which can be found in Chen and Sohal,[1] for the reduced plastic moment capacity M_{pv}, due to shear force can be derived as

$$M_{pv} = M_p\left[1 - \frac{3}{4}\left(\frac{V}{V_p}\right)^2\right] \tag{3.14}$$

where V is the shear force with a maximum capacity $V_p = \tau_y \times$ area of section.

Equation (3.14) is valid only when $\frac{V}{V_p} \leq \frac{2}{3}$. In practice, the ratio V/V_p rarely exceeds 2/3. An approximate solution for rectangular sections proposed by Drucker[2] for design can also be used:

$$M_{pv} = M_p\left[1 - \left(\frac{V}{V_p}\right)^4\right] \tag{3.15}$$

For I sections, only the web is usually assumed to resist the shear force when the magnitude of the shear force is small. For significant shear force, the bending moment will be reduced. This assumption leads to a bending moment–shear interaction relationship as plotted in Figure 3.4. Further simplification of the bending–shear relationship has been made for design code use. The Australian steel design code adopts the formula given in Equation (3.16), which is also plotted in Figure 3.4.

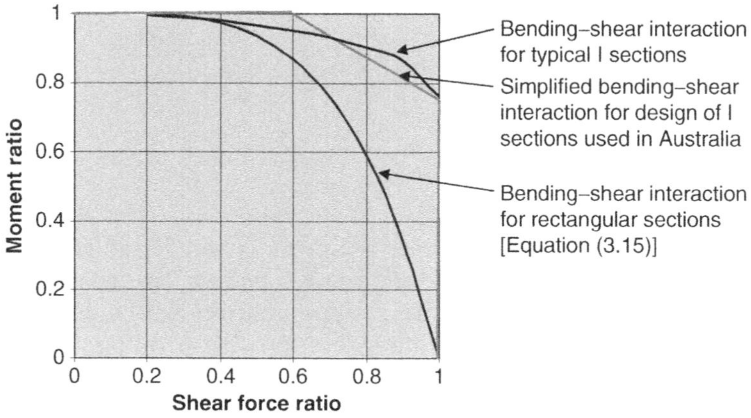

FIGURE 3.4. Bending–shear interaction.

$$M_{pv} = M_p \left[1.375 - 0.625 \left(\frac{V}{V_p} \right) \right] \quad (3.16)$$

Example 3.1 Structure ABC shown in Figure 3.5 is subjected to a load P acting at A. The column BC of rectangular section has a moment capacity of 160 kNm and a squash load of 480 kN. Determine the maximum load P that the column BC can support by assuming failure by (a) pure bending and (b) axial-bending interaction.

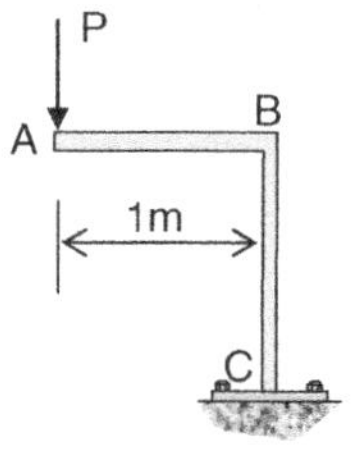

FIGURE 3.5. Example 3.1.

Solution

(a) For yielding at the fixed support C,
$P \times 1 = 160$
P = 160 kN

(b) Using the axial-bending interaction equation [Equation (3.8)] for a rectangular section,
$$\frac{P \times 1}{160} = 1 - \left(\frac{P}{480} \right)^2$$
P = 145.3 kN

3.3 Concept of Yield Surface

The variation of the bending moment with axial force in a rectangular section can be plotted in terms of the dimensionless quantities (M_{pr}/M_p) and β. The resulting curve is called the *yield surface* because any point on the yield surface represents a state of the fully yielded cross section with a bending moment, *M*, and axial force, *N*, represented by that point. For a cross section in an elastic state,

$$m = \frac{M}{M_p} \leq \frac{M_{pr}}{M_p} \quad (3.17)$$

and the forces in the cross section are represented by a point within the space bounded by the yield surface and the two axes. The yield surface of a rectangular section is shown in Figure 3.6.

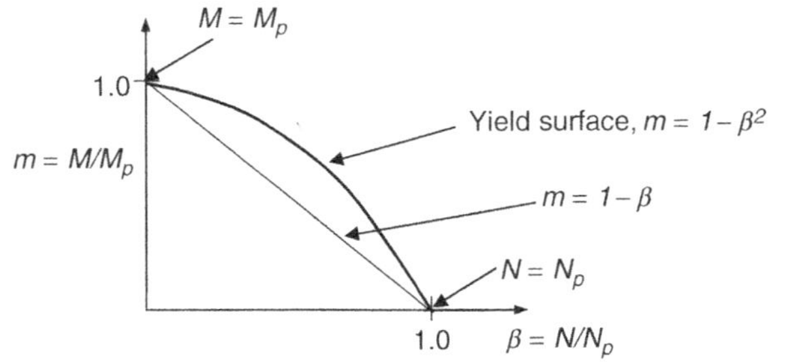

FIGURE 3.6. Yield surface of a rectangular section.

The yield surfaces for the lower and upper bounds of the practical range of I sections are plotted in Figure 3.7. Because of the complexity of Equations (3.9) and (3.11), and taking advantage of the narrow range of the yield surfaces for I sections, a simplified yield surface, represented by Equation (3.18), is usually adopted for practical design.

$$m = 1.0 \qquad \text{for } \frac{N}{N_p} \leq 0.15 \tag{3.18a}$$

$$m = 1.18(1 - \beta) \qquad \text{for } \frac{N}{N_p} > 0.15 \tag{3.18b}$$

It should be noted that the coefficient of 1.18 in Equation (3.18b) changes if the limiting axial force ratio of 0.15 changes. Some countries adopt different values for this coefficient for design. In general, the yield surface for any cross-sectional shape is concave outward as typically shown in Figures 3.6 and 3.7. The more concave the curve is, the relatively stronger the yield surface for the cross section. Therefore, a yield surface resulting from a straight line joining the

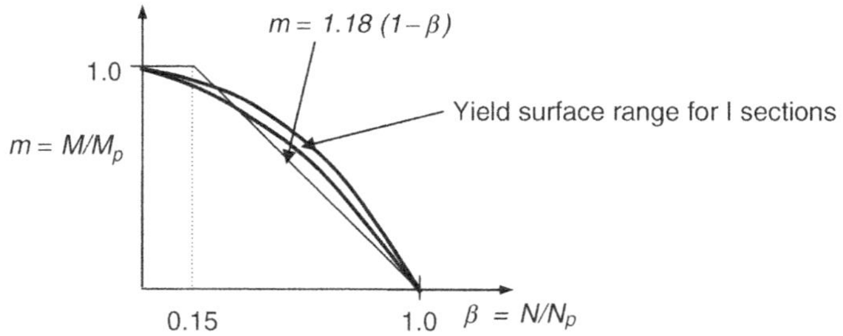

FIGURE 3.7. Yield surfaces of steel I sections.

points $m = 1.0$ and $\beta = 1.0$ represents the minimum strength of any cross section. This straight line, shown in Figure 3.6, is given by

$$m = 1 - \beta \tag{3.19}$$

For a cross section with unknown bending–axial force interaction formulation, Equation (3.19) can be used as a conservative approximation for design.

3.4 Yield Surface and Plastic Flow Rule

The yield surfaces described for various cross-sectional shapes can be presented using a yield function φ such that for a section in a fully yielded state under force interaction,

$$\phi(P_1, \ldots, P_n) = 0 \tag{3.20}$$

where $P_1, \ldots, P_n$ are the stress resultants. For a structural member in a plane frame, $n = 6$ for both ends with three stress resultants, including bending moment, shear force, and axial force at each end of the member. For example, the yield function of a rectangular section can be written as

$$\phi = m + \beta^2 - 1 \tag{3.21}$$

For a yielded section satisfying the force interaction condition, $\phi = 0$. For a section in an elastic state, $\phi < 0$. To ensure that these two conditions are always satisfied for any section, the force ratios m and β in Equation (3.21), as well as for other yield functions, are always positive when the yield surface is represented in a quadrant.

Assume that $P_1, \ldots, P_n$ specify a current state of stress resultants on the yield surface and that a change of state of stress resultants occurs by the increments $\Delta P_1, \ldots, \Delta P_n$ due to an increase in loading on the structure. Because both the original and the final states of stress resultants satisfy the yield condition [Equation (3.20)], it follows that

$$\Delta\phi = \frac{\partial\phi}{\partial P_1}\Delta P_1 + \ldots + \frac{\partial\phi}{\partial P_n}\Delta P_n = 0 \tag{3.22}$$

where the partial derivatives must be taken at the original state of stress resultants. Equation (3.22) can be written in vectorial form as

$$\Delta\phi = \{f\}^t\{\Delta P\} = 0 \tag{3.23}$$

where $\{f\}$ contains elements $f_i = \dfrac{\partial\phi}{\partial P_i}$ and $\{\Delta P\}$ represents the vector for the increments of stress resultants. The incremental force vector

$\{\Delta P\}$ is of size 6×1, corresponding to the three end forces at each end of a plane frame member. It should be noted that for ϕ containing two stress resultants, such as m and β in Equation (3.21), $\{f\}$ represents the vector containing terms related only to the bending moment and axial force. From Equation (3.23), the two vectors $\{f\}$ and $\{\Delta P\}$ must be orthogonal to each other according to vector algebra. A geometrical interpretation of the state of stress with $\{f\}$ and $\{\Delta P\}$ at point A on the yield surface is shown in Figure 3.8.

The orthogonal condition can also be applied to the relationship between the increments of stress resultants and plastic deformation as implied by Prager's[3] statement that for elastic-perfectly plastic material, "the stress increment does no work on the increment of plastic strain". When applying to frame members, this statement means that

$$\{\Delta d_p\}^t\{\Delta P\} = 0 \tag{3.24}$$

where $\{\Delta d_p\}$ is the vector of the plastic deformation increments. For example, for a section yielded under pure bending, Equation (3.24) holds because the only increment in $\{\Delta P\}$ is the bending moment, which is zero for a bilinear moment–rotation relationship. A comparison between Equations (3.23) and (3.24) indicates that $\{\Delta d_p\}$ and $\{f\}$ are acting in the same direction but not necessarily with the same magnitude. Hence, $\{\Delta d_p\}$ and $\{f\}$ can be related by

$$\{\Delta d_p\} = \lambda\{f\} \tag{3.25}$$

where λ is an arbitrary scalar factor often termed as the *plastic multiplier*. Equation (3.25) defines the plastic flow rule of structural members in frames.

It has been stated, for instance by Kaliszky[4], that for nonnegative work done by stresses going through a cycle of elastic–plastic–elastic states under changing loads, the increments in strains, and in the current case for beam elements, the plastic deformation increment vector,

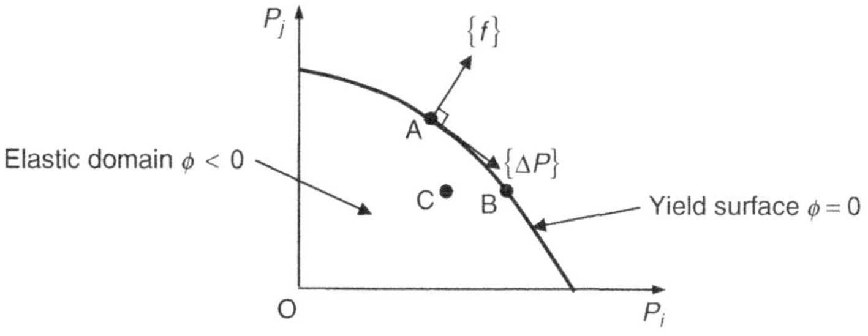

FIGURE 3.8. State of stress for a yield function.

$\{\Delta d_p\}$, must be outward normal to the tangent plane at the point on the yield surface under consideration. Hence, $\{f\}$ represents the outward normal vector to the yield surface given by Equation (3.20). This is called the *normality rule* in plasticity.

For materials in the plastic state, the plastic flow always occurs in association with a dissipation of mechanical energy. Thus, for an increment of plastic deformation $\{\Delta d_p\}$, the dissipative energy ΔW is always positive and is given by

$$\Delta W = \{P\}^t\{\Delta d_p\} = \lambda\{P\}^t\{f\} > 0 \tag{3.26}$$

where the total stress resultant vector, $\{P\}$, is represented by a line joining the origin at O (see Figure 3.8) to a point on the yield surface. Because $\{f\}$ is normal to the concave-outward yield surface, the angle between $\{P\}$ and $\{f\}$ is always less than 90°. Therefore, $\{P\}^t\{f\}$ is always positive according to vector algebra. It follows from Equation (3.26) that $\lambda > 0$ for a stress resultant vector staying on the yielding surface.

Suppose that the yield function at point A in Figure 3.8 is given as ϕ_A. For an increment of loading, the stress point moves to point B with the yield function given as ϕ_B. Because both ϕ_A and ϕ_B satisfy the yield condition, it follows that $\Delta\phi = \phi_B - \phi_A = 0$, the same as Equation (3.22). However, if the material undergoes elastic unloading such that the stress point moves from A to C for which $\phi_C < 0$, it follows that $\Delta\phi = \phi_C - \phi_A < 0$. For elastic unloading, positive work is done such that

$$\{\Delta d_p\}^t\{\Delta P\} = \lambda\{f\}^t\{\Delta P\} > 0 \tag{3.27}$$

Because $\Delta\phi = \{\Delta P\}^t\{f\} < 0$, therefore $\lambda < 0$. The sign of λ is therefore used as an indication of whether elastic unloading occurs in an incremental elastoplastic analysis.

In summary, the state of stress in a section can be classified as

Elastic state: $\lambda = 0$ and $\phi < 0$;
Plastic state to plastic state: $\lambda \geq 0$, $\phi = 0$ and $\Delta\phi = 0$;
Plastic state to elastic state (elastic unloading): $\lambda < 0$, $\phi < 0$ and $\Delta\phi < 0$.

3.4.1 Plastic Multiplier and General Elastoplastic Stiffness Matrix

For a section in plastic state, the incremental deformation vector, $\{\Delta d\}$, consists of both elastic and plastic displacements, depending on which force components are active in the yield function. For example, yielding by pure bending induces plastic rotation, whereas axial force and shear force induce respective elastic deformation. Hence,

$$\{\Delta d\} = \{\Delta d_e\} + \{\Delta d_p\} \tag{3.28}$$

where the incremental elastic displacement vector $\{\Delta d_e\}$ is related to the incremental force vector by

$$\{\Delta P\} = [K_e]\{\Delta d_e\} \tag{3.29}$$

It should be pointed out that Equation (3.29) is equivalent to Equation (1.9) in Chapter 1 in incremental form.

Substituting $\{\Delta d_e\}$ from Equation (3.28) and $\{\Delta d_p\}$ from Equation (3.25) into Equation (3.29), we obtain

$$\{\Delta P\} = [K_e]\{\{\Delta d\} - \lambda\{f\}\} \tag{3.30}$$

Using Equation (3.30) in Equation (3.23), the plastic multiplier λ can be found to be

$$\lambda = \frac{\{f\}^t[K_e]}{\{f\}^t[K_e]\{f\}}\{\Delta d\} \tag{3.31}$$

Substituting Equation (3.31) into Equation (3.30), the elastoplastic stiffness matrix, $[K_{pe}]$, can be found:

$$\{\Delta P\} = [K_{Pe}]\{\Delta d\} \tag{3.32}$$

where $[K_{Pe}] = [K] - \dfrac{[K_e]\{f\}\{f\}^t[K_e]}{\{f\}^t[K_e]\{f\}}$

Equation (3.32) is a general expression for a yielded beam element. Since a beam element may be subjected to different combinations of yielding states at its ends, the form of $[K_{Pe}]$ varies according to the state of yielding and the yield function adopted for plastic analysis.

3.5 Derivation of General Elastoplastic Stiffness Matrices

In finite element analysis, elastoplastic stiffness matrix derived in a form similar to Equation (3.32) has been commonly used in material nonlinear analysis. Ueda *et al.*[5] were among the first to apply the theory to beam elements in frame analysis. For a beam element, the plastic state at its two ends is considered and the corresponding elastoplastic stiffness matrix derived. As shown in Figure 3.9, there are four cases of yield condition for a beam element to be considered: (a) both ends *i* and *j* are elastic, (b) end *i* is plastic and end *j* is elastic, (c) end *i* is elastic and end *j* is plastic, and (d) both ends *i* and *j* are plastic. Explicit

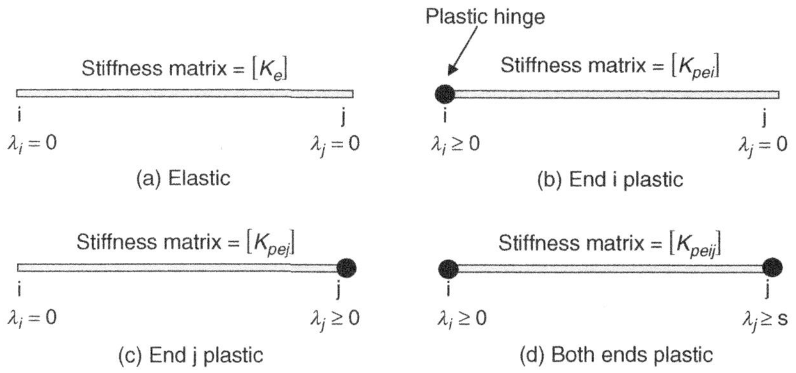

FIGURE 3.9. Four yielding cases for beam element.

expressions of $[K_{Pe}]$ for each of these four cases are given. In deriving $[K_{Pe}]$, the force and displacement vectors for each element are expressed in two parts corresponding to the two ends i and j such that Equation (3.28) can be rewritten as

$$\{\Delta d\} = \left\{ \begin{array}{c} \Delta d_i \\ \Delta d_j \end{array} \right\} = \left\{ \begin{array}{c} \Delta d_{ei} \\ \Delta d_{ej} \end{array} \right\} + \left\{ \begin{array}{c} \Delta d_{pi} \\ \Delta d_{pj} \end{array} \right\} \tag{3.33}$$

and

$$\{\Delta P\} = \left\{ \begin{array}{c} \Delta P_i \\ \Delta P_j \end{array} \right\} \tag{3.34}$$

For the incremental plastic displacement vector in Equation (3.33),

$$\left\{ \begin{array}{c} \Delta d_{pi} \\ \Delta d_{pj} \end{array} \right\} = \begin{bmatrix} f_i & 0 \\ 0 & f_j \end{bmatrix} \left\{ \begin{array}{c} \lambda_i \\ \lambda_j \end{array} \right\} \tag{3.35}$$

according to Equation (3.25). The elastic stiffness matrix can be expressed as

$$[K_e] = \begin{bmatrix} K_{ii} & K_{ij} \\ K_{ji} & K_{jj} \end{bmatrix} \tag{3.36}$$

3.5.1 Both Ends are Elastic

In this case, $\lambda_i = \lambda_j = 0$. Hence,

$$[K_{Pe}] = [K_e] \tag{3.37}$$

3.5.2 End i is Plastic and End j is Elastic

This case implies that $\lambda_j = 0$. From Equation (3.30),

$$\begin{Bmatrix} \Delta P_i \\ \Delta P_j \end{Bmatrix} = [K_e]\{\Delta d\} - \begin{bmatrix} K_{ii} \\ K_{ji} \end{bmatrix} \lambda_i \{f_i\} \tag{3.38}$$

Because $\lambda_i \{f_i\}^t \{\Delta P_i\} = 0$, therefore

$$\lambda_i = \frac{\{f_i\}^t [K_{ii} \quad K_{ij}]}{\{f_i\}^t [K_{ii}]\{f_i\}} \{\Delta d\} \tag{3.39}$$

$$[K_{Pe}] = [K_{Pei}] = [K_e] - \frac{\begin{bmatrix} K_{ii} \\ K_{ji} \end{bmatrix} \{f_i\}\{f_i\}^t [K_{ii} \quad K_{ij}]}{\{f_i\}^t [K_{ii}]\{f_i\}} \tag{3.40}$$

In the equations just given, $\{f_i\} = \begin{Bmatrix} \dfrac{\partial \phi}{\partial N_i} \\ \dfrac{\partial \phi}{\partial Q_i} \\ \dfrac{\partial \phi}{\partial M_i} \end{Bmatrix}$ for a plane frame member. If ϕ consists of only N and M as in bending-axial interaction yield function, then $\partial \phi / \partial Q_i = 0$.

3.5.3 End i is Elastic and End j is Plastic

This case implies that $\lambda_i = 0$. Using a similar procedure as given earlier, λ_j and $[K_{Pe}]$ can be derived as

$$\lambda_j = \frac{\{f_j\}^t [K_{ji} \quad K_{jj}]}{\{f_j\}^t [K_{jj}]\{f_j\}} \{\Delta d\} \tag{3.41}$$

$$[K_{Pe}] = [K_{Pej}] = [K_e] - \frac{\begin{bmatrix} K_{ij} \\ K_{jj} \end{bmatrix} \{f_j\}\{f_j\}^t [K_{ji} \quad K_{jj}]}{\{f_j\}^t [K_{jj}]\{f_j\}} \tag{3.42}$$

where $\{f_j\} = \begin{Bmatrix} \dfrac{\partial \phi}{\partial N_j} \\ \dfrac{\partial \phi}{\partial Q_j} \\ \dfrac{\partial \phi}{\partial M_j} \end{Bmatrix}$.

3.5.4 Both End i and End j are Plastic

In this case, both λ_i and λ_j need to be evaluated. By writing $[G]=\begin{bmatrix} f_i & 0 \\ 0 & f_j \end{bmatrix}$, then

$$\begin{Bmatrix} \lambda_i \\ \lambda_j \end{Bmatrix} = [G^t \quad K_e \quad G]^{-1}[G]^t[K_e]\{\Delta d\} \tag{3.43}$$

$$[K_{Pe}] = [K_{Peij}] = [K_e] - [K_e][G][G^t \quad K_e \quad G]^{-1}[G]^t[K_e] \tag{3.44}$$

3.6 Elastoplastic Stiffness Matrices for Sections

Different yield surfaces used for the beam element with different yield conditions give different expressions for the elastoplastic stiffness matrices. These expressions can be given explicitly on the basis of the derivations described in Section 3.5. In the following section, elastoplastic stiffness matrices based on the yield surfaces for I sections [Equations (3.18)] and for general sections [Equations (3.19)] are given.

3.6.1 End i is Plastic and End j is Elastic

In this case, $\{f_j\}=0$. For I sections, the yield surface consists of two planes, termed hyperplanes, to be considered. One hyperplane is where the axial force ratio $\beta \le 0.15$. The other is where $\beta > 0.15$.

Case (a): for I sections when $\beta \le 0.15$ and for any section based on yielding by pure bending [Equations (3.18a)].

$$\{f_i\} = \begin{Bmatrix} \partial\phi/\partial N_i \\ \partial\phi/\partial Q_i \\ \partial\phi/\partial M_i \end{Bmatrix} = \begin{Bmatrix} 0 \\ 0 \\ 1/M_p \end{Bmatrix} \tag{3.45}$$

Expanding Equations (3.39) and (3.40) using Equation (3.45), we get

$$\lambda_i = \frac{M_p}{2}\begin{bmatrix} 0 & 3/L & 2 & 0 & -3/L & 1 \end{bmatrix}\{\Delta \mathrm{d}\} \tag{3.46}$$

$$[K_{Pei}] = \frac{EI}{\eta L}\begin{bmatrix} 1 & 0 & 0 & -1 & 0 & 0 \\ & 3\eta/L^2 & 0 & 0 & -3\eta/L^2 & 3\eta/L \\ & & 0 & 0 & 0 & 0 \\ & & & 1 & 0 & 0 \\ & \textit{Symmetric} & & & 3\eta/L^2 & -3\eta/L \\ & & & & & 3\eta \end{bmatrix} \tag{3.47}$$

where $\eta = I/A$. The symbols in both Equations (3.46) and (3.47), and in the following equations, follow those used in Chapter 1.

Case (b): for I sections where $\gamma = 1.18\dfrac{M_p}{N_p}$ when $\beta > 0.15$ or $\gamma = \dfrac{M_p}{N_p}$ for general sections.

Writing $\{f_i\} = \dfrac{1}{M_p}\begin{Bmatrix} \gamma \\ 0 \\ 1 \end{Bmatrix}$, it can be derived that

$$\lambda_i = \frac{M_p}{\gamma^2 + 4\eta}\begin{bmatrix} \gamma & 6\eta/L & 4\eta & -\gamma & -6\eta/L & 2\eta \end{bmatrix}\{\Delta d\} \tag{3.48}$$

$$[K_{Pei}] = \frac{EI}{L(\gamma^2 + 4\eta)}\begin{bmatrix} 4 & -g & -2h & -4 & g & -h \\ & b & g\gamma & g & -b & e \\ & & h^2 & 2h & -g\gamma & h^2/2 \\ & & & 4 & -g & h \\ & \text{Symmetric} & & & b & -e \\ & & & & & c \end{bmatrix} \tag{3.49}$$

where $b = \dfrac{12(\gamma^2 + \eta)}{L^2}$, $c = 4(\gamma^2 + 3\eta)$, $e = \dfrac{6(\gamma^2 + 2\eta)}{L}$, $g = \dfrac{6\gamma}{L}$, and $h = 2\gamma$.

3.6.2 End *i* is Elastic and End *j* is Plastic

Case (a): for I sections when $\beta \leq 0.15$ and for any section based on yielding by pure bending.

In this case, $\{f_i\} = 0$,

$$\{f_i\} = \begin{Bmatrix} \partial\phi/\partial N_j \\ \partial\phi/\partial Q_j \\ \partial\phi/\partial M_j \end{Bmatrix} = \begin{Bmatrix} 0 \\ 0 \\ 1/M_p \end{Bmatrix} \tag{3.50}$$

Expanding Equations (3.41) and (3.42) using Equation (3.50), we obtain

$$\lambda_j = \frac{M_p}{2}\begin{bmatrix} 0 & 3/L & 1 & 0 & -3/L & 2 \end{bmatrix}\{\Delta d\} \tag{3.51}$$

$$[K_{pej}] = \frac{EI}{\eta L}\begin{bmatrix} 1 & 0 & 0 & -1 & 0 & 0 \\ & 3\eta/L^2 & 3\eta/L & 0 & -3\eta/L^2 & 0 \\ & & 3\eta & 0 & -3\eta/L & 0 \\ & & & 1 & 0 & 0 \\ & \text{Symmetric} & & & 3\eta/L^2 & 0 \\ & & & & & 0 \end{bmatrix} \tag{3.52}$$

Case (b): for I sections where $\gamma = 1.18\frac{M_p}{N_p}$ when $\beta > 0.15$ or $\gamma = \frac{M_p}{N_p}$ for general sections.

Writing $\{f_j\} = \frac{1}{M_p}\begin{Bmatrix} \gamma \\ 0 \\ 1 \end{Bmatrix}$, it can be derived that

$$\lambda_j = \frac{M_p}{\gamma^2 + 4\eta}\begin{bmatrix} -\gamma & 6\eta/L & 2\eta & \gamma & -6\eta/L & 4\eta \end{bmatrix}\{\Delta d\} \tag{3.53}$$

$$[K_{pej}] = \frac{EI}{L(\gamma^2 + 4\eta)}\begin{bmatrix} 4 & g & h & -4 & -g & 2h \\ & b & e & -g & -b & g\gamma \\ & & c & -h & -e & h^2/2 \\ & & & 4 & g & -2h \\ & \text{Symmetric} & & & b & -g\gamma \\ & & & & & h^2 \end{bmatrix} \tag{3.54}$$

3.6.3 Both Ends are Plastic

Case (a): for I sections when $\beta \leq 0.15$ and for any section based on yielding by pure bending.

The outward normal vector for both ends is given as

$$[G] = \begin{bmatrix} 0 & 0 \\ 0 & 0 \\ 1/M_{pi} & 0 \\ 0 & 0 \\ 0 & 0 \\ 0 & 1/M_{pj} \end{bmatrix} \tag{3.55}$$

where M_{pi} and M_{pj} are the bending moment capacities at ends i and j, respectively.

(i) End moments have the same sign.

$$\begin{Bmatrix} \lambda_i \\ \lambda_j \end{Bmatrix} = M_{pi}\begin{bmatrix} 0 & 1/L & 1 & 0 & -1/L & 0 \\ 0 & 1/L & 0 & 0 & -1/L & 1 \end{bmatrix}\{\Delta d\} \tag{3.56}$$

(ii) End moments have different signs.

$$\begin{Bmatrix} \lambda_i \\ \lambda_j \end{Bmatrix} = M_p\begin{bmatrix} 0 & 1/L & 1 & 0 & -1/L & 0 \\ 0 & -1/L & 0 & 0 & 1/L & -1 \end{bmatrix}\{\Delta d\} \tag{3.57}$$

$$[K_{Peij}] = \frac{EI}{\eta L}\begin{bmatrix} 1 & 0 & 0 & -1 & 0 & 0 \\ & 0 & 0 & 0 & 0 & 0 \\ & & 0 & 0 & 0 & 0 \\ & & & 1 & 0 & 0 \\ & \mathit{Symmetric} & & & 0 & 0 \\ & & & & & 0 \end{bmatrix} \tag{3.58}$$

Case (b): for I sections where $\gamma = 1.18\frac{M_p}{N_p}$ when $\beta > 0.15$ or $\gamma = \frac{M_p}{N_p}$ for general sections.

There are two possible forms for the elastoplastic stiffness matrix depending on the signs of the end moments.

(i) End moments have the same sign; $M_{pi} = M_{pj}$.

$$[G] = \frac{1}{M_{pi}}\begin{bmatrix} \gamma & 0 \\ 0 & 0 \\ 1 & 0 \\ 0 & -\gamma \\ 0 & 0 \\ 0 & 1 \end{bmatrix} \tag{3.59}$$

$$\begin{Bmatrix} \lambda_i \\ \lambda_j \end{Bmatrix} = \frac{M_{pi}}{2(\gamma^2 + 3\eta)}\begin{bmatrix} \gamma & 6\eta/L & \gamma^2 + 6\eta & -\gamma & -6\eta/L & -\gamma^2 \\ \gamma & 6\eta/L & -\gamma^2 & -\gamma & -6\eta/L & \gamma^2 + 6\eta \end{bmatrix} \tag{3.60}$$

$$[K_{peij}] = \frac{3EI}{L(\gamma^2 + 3\eta)}\begin{bmatrix} 1 & -2\gamma/L & -\gamma & -1 & 2\gamma/L & -\gamma \\ & 4\gamma^2/L^2 & 2\gamma^2/L & 2\gamma/L & -4\gamma^2/L^2 & 2\gamma^2/L \\ & & \gamma^2 & \gamma & -2\gamma^2/L & \gamma^2 \\ & & & 1 & -2\gamma/L & \gamma \\ & \mathit{Symmetric} & & & 4\gamma^2/L^2 & -2\gamma^2/L \\ & & & & & \gamma^2 \end{bmatrix} \tag{3.61}$$

(ii) End moments have different signs; $M_{pi} = -M_{pj}$

$$[G] = \frac{1}{M_{pi}}\begin{bmatrix} \gamma & 0 \\ 0 & 0 \\ 1 & 0 \\ 0 & -\gamma \\ 0 & 0 \\ 0 & -1 \end{bmatrix} \tag{3.62}$$

$$\begin{Bmatrix} \lambda_i \\ \lambda_j \end{Bmatrix} = \frac{M_{pi}}{2(\gamma^2+\eta)} \begin{bmatrix} \gamma & 2(\gamma^2+\eta)/L & \gamma^2+2\eta & -\gamma & -2(\gamma^2+\eta)/L & \gamma^2 \\ \gamma & -2(\gamma^2+\eta)/L & -\gamma^2 & -\gamma & 2(\gamma^2+\eta)/L & -(\gamma^2+2\eta) \end{bmatrix} \tag{3.63}$$

$$[K_{Peij}] = \frac{EI}{L(\gamma^2+\eta)} \begin{bmatrix} 1 & 0 & -\gamma & -1 & 0 & \gamma \\ & 0 & 0 & 0 & 0 & 0 \\ & & \gamma^2 & \gamma & 0 & -\gamma^2 \\ & & & 1 & 0 & -\gamma \\ & \text{Symmetric} & & & 0 & 0 \\ & & & & & \gamma^2 \end{bmatrix} \tag{3.64}$$

In the derivations just given, the sign of M_p depends on the calculation of the incremental load factors α_A, α_B, etc. Details of the determination of the sign of M_p are given in Chapter 4.

3.7 Stiffness Matrix and Elastoplastic Analysis

Use of the stiffness matrices derived in Section 3.6 in an incremental elastoplastic analysis represents a direct satisfaction of the yield conditions for any plastic hinge formed during the analysis. The analysis gives information on the increment of forces, as well as displacements for both elastic and plastic components. However, the information on the plastic displacements generated by the plastic hinges is given implicitly in the results. That is, the solution of the equilibrium equation of the structure yields only the total displacement vector $\{\Delta d\}$. The plastic rotation, for instance, of a plastic hinge must be calculated separately using Equation (3.25). Hence, simulation of plastic hinge behavior in an incremental elastoplastic analysis is similar to the condensation method for modeling the behavior of a pin as described in Section 1.12.2 of Chapter 1.

Indeed, the stiffness matrices given in Equations (3.47), (3.52), and (3.58) for cases of yielding by pure bending are identical to Equations (1.32), (1.39), and (1.46), respectively, at the corresponding ends with a pin. For plastic design where limited plastic deformation capacity is a design criterion, separate calculations of the plastic deformation vector are necessary.

The following example demonstrates the difference between using the pure bending yield criterion and the bending-axial interaction yield criterion. The effects of these yield criteria on the collapse load of the structure are also shown. Only the results of the analyses are given. Details of the elastoplastic analysis are explained in the following chapter.

Example 3.2 Determine the elastoplastic collapse behavior of the steel frame ABC shown in Figure 3.10. Member 1 is fixed to a rigid wall at A while member 2 is fixed to a roller support at C. Yield criteria for both pure bending and I sections bending about the strong axis [Equations (3.18)] are used for structural members that have the following properties: $E = 2 \times 10^8$ kN/m^2, $A = 0.0105$ m^2, $I = 0.000477$ m^4, $M_p = 515$ kNm.

Solution. Load–deflection curves showing variation of the load factor α with the horizontal deflection at C for both cases are shown in Figure 3.11. The corresponding values of load factors and deflections are given in Table 3.1. The sequence of plastic hinge formation for both cases is (1) hinge at C of member 2, (2) hinge at B of member 2, and (3) hinge at A of member 1. The effect of the axial-bending interaction yield criterion reduces the collapse load factor of the structure from 1.288 for the pure bending case to 1.069, or by 17%.

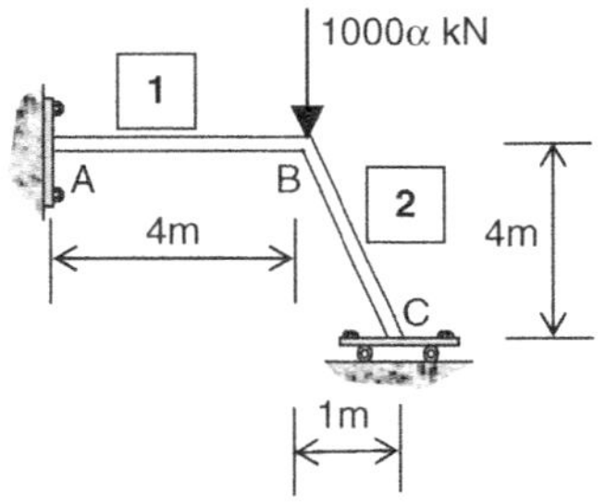

FIGURE 3.10. Example 3.2.

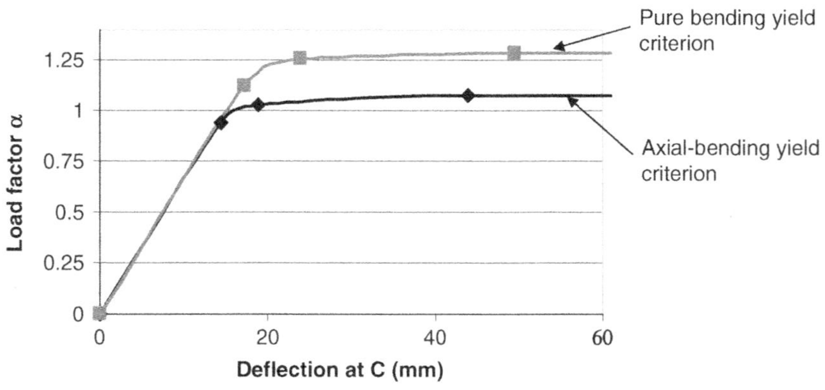

FIGURE 3.11. Load–deflection curves.

TABLE 3.1
Results of load–deflection for Example 3

Pure bending case		*Axial-bending interaction case*	
Load factor α	*Deflection at C*	*Load factor* α	*Deflection at C*
1.118	14.4	0.935	17.2
1.259	18.9	1.023	23.9
1.288	43.9	1.069	49.5

3.8 Modified End Actions

It has been shown in Chapter 1 that when using the condensation method for modeling pins at the ends of a member subjected to internal loads, the resulting fixed-end forces vary according to the end conditions. The condensation method is based on modeling of the pins implicitly by modifying the member stiffness matrix. For example, if a member includes a pin at one end and is fixed at the other, the modified fixed-end force vector is derived as equivalent to that of a propped cantilever beam.

In elastoplastic analysis, where yield is accounted for by pure bending and the resulting plastic hinge is modeled by a pin, the derivation of the elastoplastic stiffness matrices in the previous sections of this chapter results in modified fixed-end force vectors identical to those derived in Chapter 1. However, for yield being accounted for by force interactions, the modified fixed-end force vectors in an elastoplastic analysis, hereafter termed "modified end actions" vectors to distinguish them from those used in linear elastic analysis, can be derived in a similar way.

In the presence of internal loads such as distributed loads, temperature effects, and support settlement, the equilibrium of a member with fixed ends in an elastic state can be written in incremental form as

$$\{\Delta P\} = [K_e]\{\Delta d_e\} + \{\Delta P_F\} \tag{3.65}$$

where $\{\Delta P_F\}$ is the fixed-end force vector. For members with internal loads, Equation (3.65) should be used instead of Equation (3.29) to derive the elastoplastic stiffness matrices when the members become yielded. The corresponding equilibrium equation of the yielded member can be written as

$$\{\Delta P\} = [K_{Pe}]\{\Delta d\} + \{\Delta P^M\} \tag{3.66}$$

where $\{\Delta P^M\}$ is the modified end action vector. Again, there are different forms of $\{\Delta P^M\}$ according to the plastic states of the member.

To derive generic and explicit expressions for $\{\Delta P^M\}$, the elastic fixed-end force vector is written as

$$\{\Delta P_F\} = \begin{Bmatrix} \Delta P_{Fi} \\ \Delta P_{Fj} \end{Bmatrix} = \begin{Bmatrix} N_i \\ S_i \\ M_i \\ N_j \\ S_j \\ M_j \end{Bmatrix} \tag{3.67}$$

and the general outward normal vector as

$$\{f_i\} = \{f_j\} = \frac{1}{M_p}\begin{Bmatrix} \gamma \\ 0 \\ 1 \end{Bmatrix}$$

3.8.1 Both Ends are Elastic

In this case, $\{\Delta d\} = \{\Delta d_e\}$ and $[K_{Pe}] = [K_e]$. Hence,

$$\{\Delta P^M\} = \{\Delta P_F\} \tag{3.68}$$

3.8.2 End i is Plastic and End j is Elastic

Following the same derivation process as in Section 3.5, the plastic multiplier and the modified end action vector are derived as

$$\lambda_i = \frac{\{f_i\}^t[K_{ii}K_{ij}]\{\Delta d\} + \{f_i\}^t\{\Delta P_{Fi}\}}{\{f_i\}^t[K_{ii}]\{f_i\}} \tag{3.69}$$

$$\{\Delta P_i^M\} = \{\Delta P_F\} - \frac{\begin{bmatrix} K_{ii} \\ K_{ji} \end{bmatrix}\{f_i\}\{f_i\}^t\{\Delta P_{Fi}\}}{\{f_i\}^t[K_{ii}]\{f_i\}}$$

$$= \frac{1}{\gamma^2 + 4\eta}\begin{Bmatrix} 4\eta N_i - \gamma M_i \\ S_i(\gamma^2 + 4\eta) - 6\eta(\gamma N_i + M_i)/L \\ \gamma^2 M_i - 4\gamma\eta N_i \\ N_j(\gamma^2 + 4\eta) + \gamma^2 N_i + \gamma M_i \\ S_j(\gamma^2 + 4\eta) + 6\eta(\gamma N_i + M_i)/L \\ M_j(\gamma^2 + 4\eta) - 2\eta(\gamma N_i + M_i) \end{Bmatrix} \tag{3.70}$$

3.8.3 End i is Elastic and End j is Plastic

$$\lambda_j = \frac{\{f_j\}^t[K_{ji} \quad K_{jj}]\{\Delta d\} + \{f_j\}^t\{\Delta P_{Fj}\}}{\{f_j\}^t[K_{jj}]\{f_j\}} \tag{3.71}$$

$$\left\{\Delta P_j^M\right\} = \{\Delta P_F\} - \frac{\begin{bmatrix} K_{ij} \\ K_{jj} \end{bmatrix}\{f_j\}\{f_j\}^t\{\Delta P_{Fj}\}}{\{f_j\}^t[K_{jj}]\{f_j\}}$$

$$= \frac{1}{\gamma^2 + 4\eta}\left\{\begin{array}{c} N_i(\gamma^2 + 4\eta) + \gamma^2 N_j + \gamma M_j \\ S_i(\gamma^2 + 4\eta) - 6\eta(\gamma N_j + M_j)/L \\ M_i(\gamma^2 + 4\eta) - 2\eta(\gamma N_j + M_j) \\ 4\eta N_j - \gamma M_j \\ S_j(\gamma^2 + 4\eta) + 6\eta(\gamma N_j + M_j)/L \\ \gamma^2 M_j - 4\gamma\eta N_j \end{array}\right\} \tag{3.72}$$

3.8.4 Both End i and End j are Plastic

Again, two separate cases are considered, depending on the signs of the end moments.

(i) End moments have the same sign; $M_{pi} = M_{pj}$.
For [G] given as in Equations (3.59), it can be derived that

$$\begin{Bmatrix} \lambda_i \\ \lambda_j \end{Bmatrix} = \frac{[G]^t[K_e]\{\Delta d\} + [G]^t\{\Delta P_F\}}{[G]^t[K_e][G]} \tag{3.73}$$

$$\left\{\Delta P_{ij}^M\right\} = \{\Delta P_F\} - \frac{[K_e][G][G]^t\{\Delta P_F\}}{[G]^t[K_e][G]}$$

$$= \frac{1}{2(\gamma^2 + 3\eta)}\left\{\begin{array}{c} 2N_i(\gamma^2 + 3\eta) - \gamma(\gamma N_i + M_i - \gamma N_j + M_j) \\ 2S_i(\gamma^2 + 3\eta) - 6\eta(\gamma N_i + M_i - \gamma N_j + M_j)/L \\ 2M_i(\gamma^2 + 3\eta) - [\gamma(\gamma^2 + 6\eta)N_i + (\gamma^2 + 6\eta)M_i + \gamma^3 N_j - \gamma^2 M_j] \\ 2N_j(\gamma^2 + 3\eta) + \gamma(\gamma N_i + M_i - \gamma N_j + M_j) \\ 2S_j(\gamma^2 + 3\eta) + 6\eta(\gamma N_i + M_i - \gamma N_j + M_j)/L \\ 2M_j(\gamma^2 + 3\eta) + [\gamma^3 N_i + \gamma^2 M_i + \gamma(\gamma^2 + 6\eta)N_j - (\gamma^2 + 6\eta)M_j] \end{array}\right\} \tag{3.74}$$

(ii) End moments have different signs; $M_{pi} = M_{pj}$.
For $[G]$ given as in Equation (3.62), it can be derived that

$$\left\{\Delta P_{ij}^{M}\right\} = \frac{1}{2(\gamma^2+\eta)}\left\{\begin{array}{c} 2N_i(\gamma^2+\eta) - \gamma(\gamma N_i + M_i - \gamma N_j - M_j) \\ 2S_i(\gamma^2+\eta) - 2[\gamma(\gamma^2+\eta)(N_i+N_j) + (\gamma^2+\eta)(M_i+M_j)]/L \\ 2M_i(\gamma^2+\eta) - [\gamma(\gamma^2+2\eta)N_i + (\gamma^2+2\eta)M_i + \gamma^3 N_j + \gamma^2 M_j] \\ 2N_j(\gamma^2+\eta) + \gamma(\gamma N_i + M_i - \gamma N_j - M_j) \\ 2S_j(\gamma^2+\eta) + 2[\gamma(\gamma^2+\eta)(N_i+N_j) + (\gamma^2+\eta)(M_i+M_j)]/L \\ 2M_j(\gamma^2+\eta) - [\gamma^3 N_i + \gamma^2 M_i + \gamma(\gamma^2+2\eta)N_j + (\gamma^2+2\eta)M_j] \end{array}\right\} \tag{3.75}$$

3.9 Linearized Yield Surface

The normal vector $\{f\}$ on the yield surface is derived by partial differentiation of the yield function ϕ. Therefore, for a linear yield surface represented by a straight line, $\{f\}$ contains only constant terms. For a nonlinear yield surface, such as that of a rectangular section, $\{f\}$ contains terms of the current forces. Because $[K_{Pe}]$ is calculated in terms of $\{f\}$ and $[K_e]$, which is a constant for linear elastic elements, $[K_{Pe}]$ is a constant if $\{f\}$ contains only constant terms. Similarly, $[K_{Pe}]$ is nonlinear if $\{f\}$ contains current forces for nonlinear yield surfaces.

Consider a nonlinear yield surface as shown in Figure 3.12a. Because $\{f\}$ and $\{\Delta P\}$ are orthogonal to each other, the increment in forces of a yielded element will invariably cause violation of the yield condition as the force vector $\{\Delta P\}$ drifts from A to B. Methods to

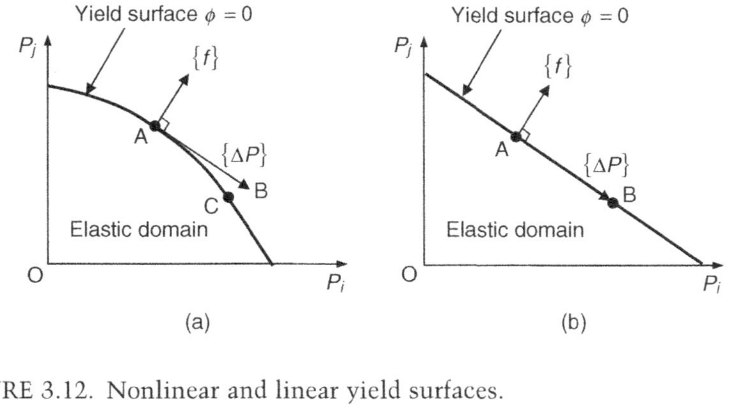

FIGURE 3.12. Nonlinear and linear yield surfaces.

restore satisfaction of the yield condition have been proposed by some researchers (e.g., Orbison *et al.*[6]). The restoration scheme is usually performed by artificially bringing the force vector point from B to C on the yield surface. The difficulty in this scheme is that it is hard to determine the exact location of C. To minimize error, the incremental step of the analysis is set to be small so that points B and C are close to each other.

For a linear yield surface as shown in Figure 3.12b, the incremental force vector $\{\Delta P\}$ of a yielded element drifts along the yield surface so that the yield condition is always satisfied. Therefore, it may be advantageous in computation to transform a nonlinear yield surface into a series of linear ones. By doing so, the accuracy of the solution depends only on the number of linearized hyperplanes representing the original nonlinear yield surface. Such a scheme has been implemented in Wong and Tin-Loi[7] and is commonly adopted in mathematical programming for elastoplastic analysis (see Chapter 6).

Problems

3.1. A structure ABC shown in Figure P3.1 is subjected to a load P acting at A. The column BC of rectangular section has a moment capacity of 160 kNm and a squash load of 480 kN. Determine the maximum load P that the column BC can support by assuming failure by (a) pure bending and (b) axial-bending interaction.

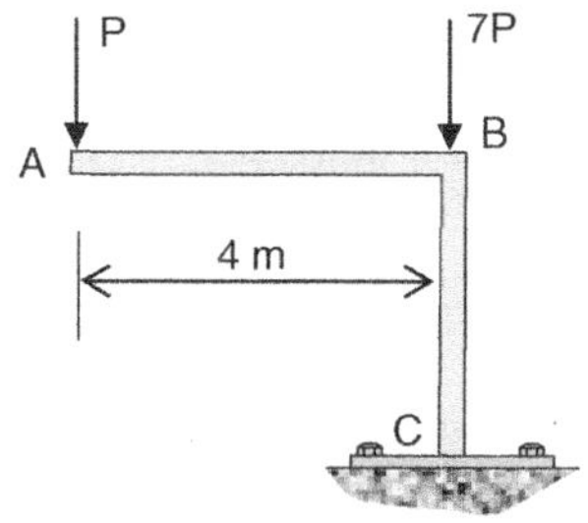

FIGURE P3.1. Problem 3.1.

3.2. The simply supported frame ABC shown in Figure P3.2 is made of I sections with a moment capacity of 240 kNm and an axial capacity of 600 kN. Determine the maximum load factor α that

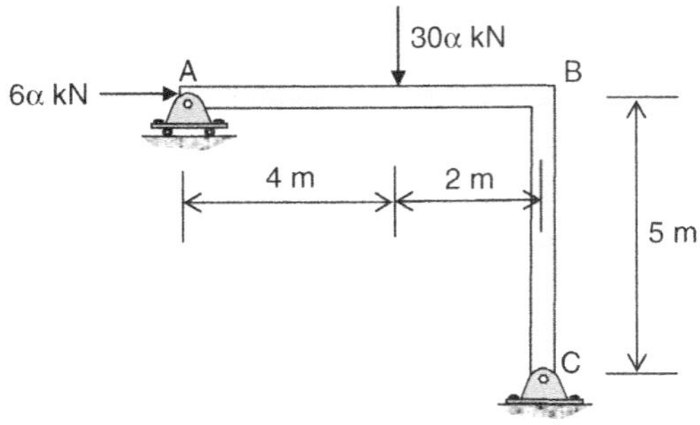

FIGURE P3.2. Problem 3.2.

the column BC can support by assuming failure by (a) pure bending and (b) axial-bending interaction.

Bibliography

1. Chen, W. F., and Sohal, I. (1995). *Plastic design and second-order analysis of steel frames*. New York: Springer-Verlag.
2. Drucker, D. C. (1956). The effect of shear on the plastic bending of beams. *J. Appl. Mech.*, **23**, pp. 509–514.
3. Prager, W. (1959). *An introduction to plasticity*, London. Addison-Wesley Pub. Co., Inc.
4. Kaliszky, S. (1989). *Plasticity: Theory and engineering applications*. New York: Elsevier.
5. Ueda, Y., Yamakawa, T., Akamatsu, T., and Matsuishi, M. (1969). A new theory on elastic-plastic analysis of framed structures. *Tech. Reports*, Osaka University, **19**, pp. 263–275.
6. Orbison, J. G., McGuire, W., and Abel, J. F. (1982). Yield surface applications in nonlinear steel frame analysis. *Comp. Meth. Appl. Mech. Eng.*, **33**, pp. 557–573.
7. Wong, M. B., and Tin-Loi, F. (1987). Yield surface linearization in elastoplastic analysis. *Comp.* and *St.*, **26**, pp. 951–956.

CHAPTER 4

Incremental Elastoplastic Analysis—Hinge by Hinge Method

4.1 Introduction

This chapter describes a method for incremental elastoplastic analysis. This method gives a complete load–deflection history of the structure until collapse. It is based on the plastic hinge concept for fully plastic cross sections in a structure under increasing proportional loading. Proportional loading applies to a structure with loads multiplied by a common load factor. This common load factor, first introduced in Chapter 3 as α (with respective values α_1, α_2, etc. at load levels 1, 2, etc.) in an incremental elastoplastic analysis, is assumed to increase until the structure collapses. The method consists of a series of elastic analyses, each of which represents the formation of a plastic hinge in the structure. Results for each elastic analysis are transferred to a spreadsheet from which the location for the formation of a plastic hinge and the corresponding increment of loading in terms of the common load factor can be obtained.

For analysis which includes the effect of force interaction on the plastification of a cross section, the method for using elastoplastic stiffness matrices for the four cases of yield condition for a beam element is applied. In contrast, an iterative method using successive approximation, with plastic hinges resulting only from yielding by pure bending, is used as an alternative to include the effect of force interaction in an incremental elastoplastic analysis. These two methods have been described in Wong[1] in which examples with detailed numerical work were given.

4.2 Use of Computers for Elastoplastic Analysis

The advent of computers has made structural analysis a good deal easier. Nowadays, commercial computer software is commonly available for linear and geometrical nonlinear elastic analyses of structural frames. Incremental elastoplastic analysis can be considered as a series of linear elastic analyses performed incrementally, perhaps using commercial computer software, in a step-by-step manner to predict plastic hinge formation. This technique, in conjunction with spreadsheet technology, can be used for elastoplastic analysis of large and complex structures for which member forces and deflections at any level of loading can be found.

As described in Chapter 3, at a load level being applied to a structure under a nominal load vector $\{F\}$, the common load factor corresponding to the formation of a plastic hinge is α_1. The solution of Equation (3.1) for the displacement increment vector $\{\Delta D\}_1$ is given as

$$\{\Delta D\}_1 = [K]^{-1}\alpha_1\{F\} \tag{4.1}$$

Because $\{F\}$ is directly proportional to the member forces, an increase in $\{F\}$ by a common load factor α_1 implies the same level of increase in the member forces. Hence, if $\{\Delta P\}$ is the vector containing the member forces for a structure under loading given by $\{F\}$, member forces for the same structure under loading given by $\alpha_1\{F\}$ must be

$$\{\Delta P\}_1 = \alpha_1\{\Delta P\} \tag{4.2}$$

If the bending moment in $\{\Delta P\}$ is M_o for an arbitrary section in a structure and the corresponding bending moment in $\{\Delta P\}_1$ is M, then for a plastic hinge to occur at the section under the pure bending yield criterion, M must be equal to M_p, the plastic moment capacity of the section. Hence, the value of α_1 leading to the formation of the plastic hinge is

$$\alpha_1 = \frac{M_p}{M_o} \tag{4.3}$$

At the load level where the load vector is $\alpha_1\{F\}$, member forces in the structure can be calculated by Equation (4.2). For any other sections not yet yielded, the remaining plastic moment capacity, M_{r1}, is generally given by

$$M_{r1} = M_p - \alpha_1 M_{o1} \tag{4.4}$$

where M_{o1} is the bending moment obtained from $\{\Delta P\}$. The remaining plastic moment capacities for all sections are then used for predicting the formation of the next plastic hinge once the yielded section has been modeled as a hinge and the structure modified according to the methods described in Chapter 3. A propped cantilever beam is used to illustrate this procedure.

Example 4.1 A propped cantilever beam is subjected to an increasing load αP shown in Figure 4.1 where P = 10 kN, L = 12 m, $M_p = 27$ kNm for the beam. What is α at collapse?

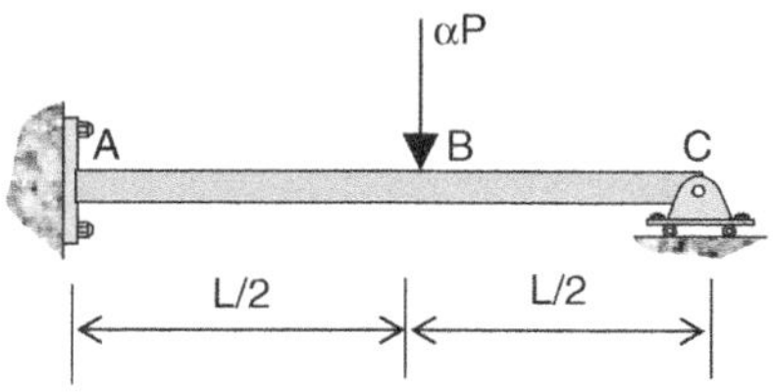

FIGURE 4.1. Elastoplastic analysis of a propped cantilever beam.

Solution. It is obvious that two plastic hinges need to occur at A and B to induce a plastic collapse mechanism. Therefore, two stages of elastic analysis are needed for the calculation of α at collapse. In addition, the vertical deflection at B can be calculated as α is increasing. Initial analysis can be carried out by assuming any value for α. For convenience, α is usually set as 1.

Stage 1:
For P = 10 kN, bending moment at A, $M_{A1} = 3PL/16 = 22.5$ kNm, and at B, $M_{B1} = 5PL/32 = 18.75$ kNm. Vertical deflection at B, $v_{B1} = 7PL^3/768EI = 157.5/EI$.

For the section at A to become plastic, from Equation (4.3), $\alpha_{A1} = \dfrac{M_p}{M_o} = \dfrac{27}{22.5} = 1.2$.

Likewise, for the section at B to become plastic, $\alpha_{B1} = \dfrac{M_p}{M_o} = \dfrac{27}{18.75} = 1.44$.

Hence, the first plastic hinge occurs at A with $\alpha_1 = 1.2$. At this load level (for a total load of $1.2 \times 10 = 12$ kN), the total bending moments at A and B are $1.2 \times 22.5 = 27$ and $1.2 \times 18.75 = 22.5$ kNm, respectively. Total vertical deflection at B is $1.2 \times 157.5/EI = 189/EI$.

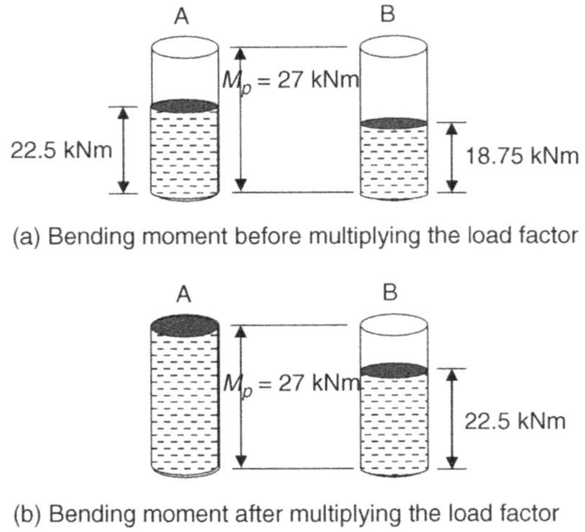

FIGURE 4.2. Cup-filling analogy for plastic hinge formation.

The aforementioned calculations for bending moments can be illustrated using a cup-filling analogy as shown in Figure 4.2.

In Figure 4.2, the height of the two empty cups represents the full plastic moment capacities at A and B. Upon the application of nominal load P, the bending moments at A and B are represented by the partially filled cups as shown in Figure 4.2a. The goal of this stage of calculation is to find a common minimum load factor α_1 so that, by multiplying the bending moments shown in Figure 4.2a by α_1, one of the cups is full, representing the occurrence of a plastic hinge. This is shown in Figure 4.2b in which the remaining plastic moment capacity at B is shown to be 4.5 kNm (= 27-22.5).

Stage 2:

At this stage of calculation, the plastic hinge at A is modeled as a real hinge and the structure ABC becomes a simply supported beam as shown in Figure 4.3.

For P = 10 kN, the increment in bending moment at B, $M_{B2} = PL/4 =$ 30 kNm and $v_{B2} = PL^3/48EI = 360/EI$.

For the section at B to become plastic, $\alpha_{B2} = \dfrac{M_p - \alpha_1 M_{B1}}{M_o} = \dfrac{4.5}{30} = 0.15.$

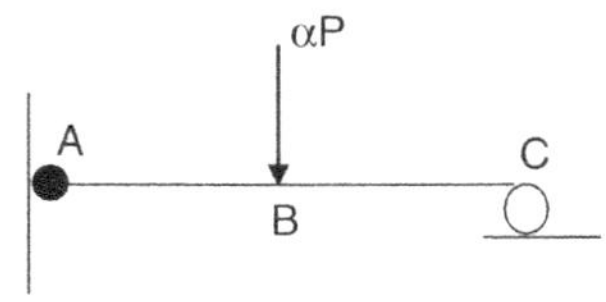

FIGURE 4.3. Stage 2 calculation.

Hence, $\alpha_2 = 0.15$. Total $M_B = 22.5 + 0.15 \times 30 = 27$ kNm. Total vertical deflection at B, $v_B = 189/EI + 0.15 \times 360/EI = 243/EI$.

Subsequent analysis after the insertion of a plastic hinge at B shows no solution or infinite deflections, indicating that the beam has reached a collapse mechanism.

Total load factor at collapse, $\alpha_{col} = \alpha_1 + \alpha_2 = 1.2 + 0.15 = 1.35$, and the collapse load is $1.35 \times 10 = 13.5$ kN.

A plot of the load factor against the vertical deflection at B is shown in Figure 4.4. In general, the gradual reduction in the slope of the load–deflection curve with increasing number of plastic hinges is an indication of the deterioration of the stiffness of the structure.

From the example just given, it is obvious that plastic hinge formation at a section corresponds to the lowest value of load factors calculated for all sections at each stage of calculation. This provides a means to identify the locations of plastic hinges without needing to guess their locations as in traditional plastic analysis. It is noted that the procedure for calculating the load factor for each of the sections and for locating the plastic hinges is routine at each loading stage. This routine procedure enables calculations to be performed in an automated way using tools such as spreadsheets. In the following section, a spreadsheet procedure is set up to carry out the routine calculations for each loading stage of the aforementioned example.

It should be noted that within each loading stage, the analysis is linear elastic. Therefore, for simple structures, standard formulas can be used for force and deflection calculations. For complex structures, forces and deflections can be calculated using commercial structural analysis computer programs, which are calculation tools commonly available in most structural engineering design offices.

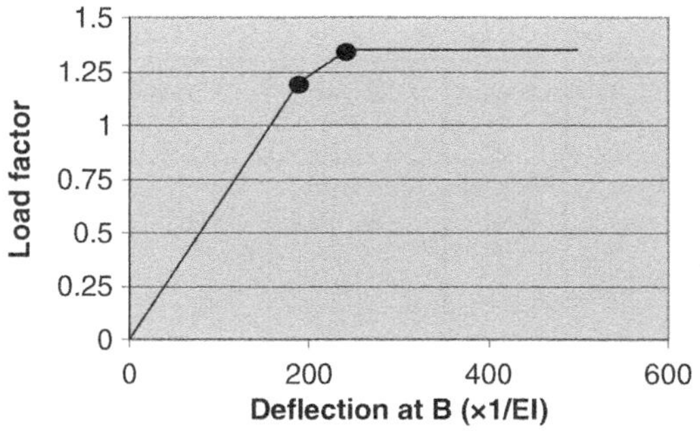

FIGURE 4.4. Load–deflection curve of a propped cantilever beam.

For structural analysis by computer, a collapse mechanism is detected when the determinant of the structure stiffness matrix is zero. In this case, a "run-time error" is usually signaled by the computer. However, in many cases, an exact zero value for the determinant of the structure stiffness matrix is difficult to detect and, instead, large deflections are obtained in the solution when the structure has reached a collapse mechanism.

4.3 Use of Spreadsheet for Automated Analysis

Steps for the routine procedure implemented on a spreadsheet for elastoplastic analysis of general structures are as described.

1. Set up a calculation table on a spreadsheet (e.g., Microsoft Excel) with headings as shown in Figure 4.5.
2. Perform a linear elastic analysis for the structure subjected to original loading (with any load factor set as 1); enter the values of bending moments M_o (Column 3) and deflections v_o (Column 7) from the results of analysis.
3. Calculate the load factor α (Column 5) for each member from Equation (4.3) such that plastic moment is reached at the ends of the member.
4. Choose the smallest load factor α_{cr} and calculate the cumulative bending moments (Column 6) and deflections (Column 8) using α_{cr} for all members. For the analysis stage i, $\alpha_i = \alpha_{cr}$ and the values of the bending moment M_i and deflection v_i are both zero when $i = 1$.
5. Calculate the residual plastic moments for all other sections (Column 4); insert a hinge in the structure at the section where α_{cr} is obtained.
6. Repeat steps 2 to 5 until the structure collapses.

Column 1	Column 2	Column 3	Column 4	Column 5	Column 6	Column 7	Column 8
FIGURE 4.5	Joint	Moment M_O	Residual plastic moment $M_P - M_i$	Load factor $\alpha = \frac{M_p - M_i}{M_o}$	Cumulative moment $M_{i+1} = M_i + \alpha_{cr} M_O$	Deflection v_O	Cumulative deflection $V_{i+1} = V_i + \alpha_{cr} V_O$

FIGURE 4.5. Spreadsheet table for elastoplastic analysis.

	A	B	C	D	E	F	G	H	I	J
1	**Example: Propped cantilever beam**									
2										
3	Stage No: 1		Critical load factor, α_{cr} =	1.2			×1/EI	×1/EI	$M_1 = 0$	
4	**Member**	**Joint**	**Moment**	**Residual Plastic Moment**	**Load factor**	**Cumulative Moment**	**Deflection**	**Cumulative deflection**	**Plastic moment**	**Mp value positive**
5			M_o	$(M_p - M_i)$	$\alpha = (M_p - M)/M_o$	$M_{i+} = M_i + \alpha_{cr} M_o$	v_o	$v_{i+1} = v_i + \alpha_{cr} v_o$	M_p	
6	1	A	22.5	27	1.2	27.0			27	27
7		B	18.75	27	1.44	22.5	157.5	189	27	27
8	2	B	-18.75	-27	1.44	-22.5			-27	27
9		C	0	27	-	0.0			27	27
10										
11	Stage No: 2		Critical load factor, α_{cr} =	0.15			×1/EI	×1/EI		
12	**Member**	**Joint**	**Moment**	**Residual Plastic Moment**	**Load factor**	**Cumulative Moment**	**Deflection**	**Cumulative deflection**	**Plastic moment**	**Mp value positive**
13			M_o	$(M_p - M_i)$	$\alpha = (M_p - M)/M_o$	$M_{i+} = M_i + \alpha_{cr} M_o$	v_o	$v_{i+1} = v_i + \alpha_{cr} v_o$	M_p	
14	1	A	0	0	-	27.0			27	27
15		B	30	4.5	0.15	27.0	360	243	27	27
16	2	B	-30	-4.5	0.15	-27.0			-27	27
17		C	0	27	-	0.0			27	27
18										
19	Total collapse load factor = 1.35									
20										

FIGURE 4.6. Spreadsheet for the elastoplastic analysis of a propped cantilever beam.

7. Theoretical collapse criterion: Determinant of structure stiffness matrix $[K] = 0$. When using computers, the collapse mechanism is reached when
   - Run-time error occurs due to zero determinant
   - Dramatic increase in displacements occurs

The final collapse load factor α_{col} is the sum of the load factors α_{cr} from all stages of analysis. Although the table in Figure 4.5 is set up for calculating bending moments and deflections, the table can be extended to include calculations of axial forces and shear forces, if desired.

A complete spreadsheet procedure using Excel for the calculation of the collapse load of the propped cantilever beam in Example 4.1 is shown in Figure 4.6. The setup of the spreadsheet is self-explanatory. The total collapse load factor in cell D19 is the sum of the cells E3 and E11. It is noted that the plastic moment M_p carries a sign that is the same as that for M_o for each of the sections. According to this rule, the sign of the plastic moment for each section is assigned under column I while column J gives its absolute value. For structures with more stages of calculations, the lines for each stage can be repeated using the copy and paste functions in Excel.

Example 4.2 Determine the collapse load factor α of the portal frame shown in Figure 4.7. $M_p = 30$ kNm for the columns and $M_p = 20$ kNm for the rafter. $E = 2 \times 10^8$ kN/m^2, $I = 0.0002$ m^4, and $A = 0.015$ m^2.

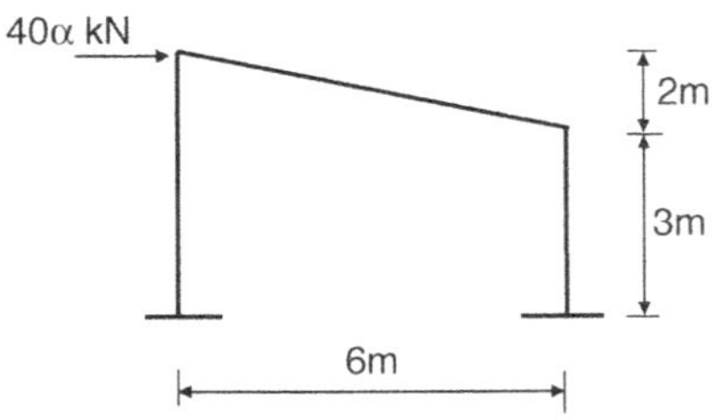

FIGURE 4.7. Elastoplastic analysis of a portal frame.

Solution. The results of analysis and the corresponding spreadsheet table for each stage of loading are as follow. The location of the plastic hinge inserted at the end of each calculation stage is also indicated. Only bending moments at the joints are calculated. In the tables, results for M_o are obtained from linear elastic analysis for a load of 40 kN. It is important to emphasize that M_p and M_o always have the same sign.

Analysis Stage No: 1 Critical Load Factor, $\alpha_{cr} = 0.503$

Member	*Joint*	*Moment* M_o	*Residual plastic moment* $M_P - M_i$	*Load factor* $\alpha = \frac{M_p - M_i}{M_o}$	*Cumulative moment* $M_{i+1} = M_i + \alpha_{cr} M_o$	*Plastic moment* M_p
1	1	27.21	30	1.103	13.69	30
	2	−21.75	−30	1.379	−10.94	−30
2	2	−21.75	−20	0.920	−10.94	−20
	3	31.00	20	0.645	15.60	20
3	3	31.00	30	0.968	15.60	30
	4	−59.62	−30	0.503	−30	−30

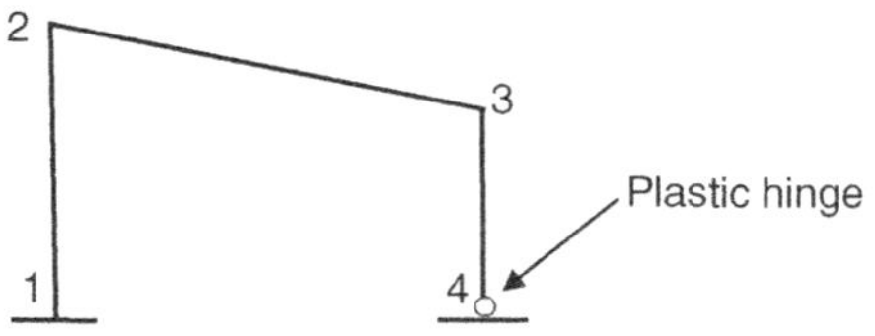

FIGURE 4.8. Results of calculations and plastic hinge location for stage 1.

Analysis Stage No: 2 Critical Load Factor, $\alpha_{cr} = 0.084$

Member	*Joint*	*Moment* M_o	*Residual plastic moment* $M_P - M_i$	*Load factor* $\alpha = \frac{M_p - M_i}{M_o}$	*Cumulative moment* $M_{i+1} = M_i + \alpha_{cr} M_o$	*Plastic moment* M_p
1	1	65.09	16.31	0.251	19.13	30
	2	−47.24	−19.06	0.403	−14.89	−30
2	2	−47.24	−9.06	0.193	−14.89	−20
	3	52.61	4.40	0.084	20	20
3	3	52.61	14.40	0.274	20	30
	4	0	0	—	−30	−30

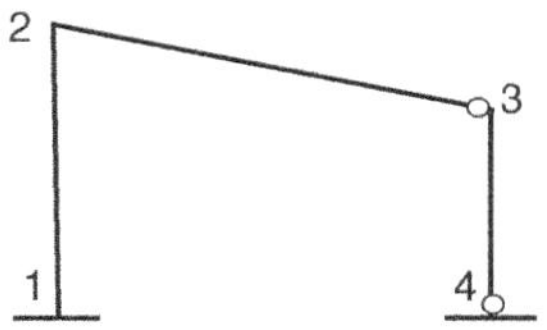

FIGURE 4.9. Results of calculations and plastic hinge location for stage 2.

Analysis Stage No: 3 Critical Load Factor, $\alpha_{cr} = 0.073$

Member	*Joint*	*Moment* M_o	*Residual plastic moment* $M_P - M_i$	*Load factor* $\alpha = \frac{M_p - M_i}{M_o}$	*Cumulative moment* $M_{i+1} = M_i + \alpha_{cr} M_o$	*Plastic moment* M_p
1	1	129.7	10.87	0.084	28.56	30
	2	−70.31	−15.11	0.215	−20	−30
2	2	−70.31	−5.11	0.073	−20	−20
	3	0	0	—	20	20
3	3	0	0	—	20	30
	4	0	0	—	−30	−30

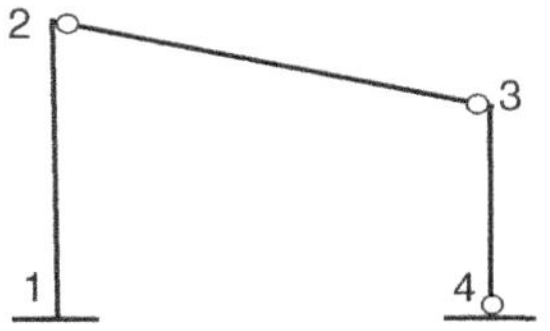

FIGURE 4.10. Results of calculations and plastic hinge location for stage 3.

Analysis Stage No: 4 Critical Load Factor, $\alpha_{cr} = 0.0072$

Member	*Joint*	*Moment* M_o	*Residual plastic moment* $M_P - M_i$	*Load factor* $\alpha = \frac{M_p - M_i}{M_o}$	*Cumulative moment* $M_{i+1} = M_i + \alpha_{cr} M_o$	*Plastic moment* M_p
1	1	200	1.44	0.0072	30	30
	2	0	−10	—	−20	−30
2	2	0	0	—	−20	−20
	3	0	0	—	20	20
3	3	0	0	—	20	30
	4	0	0	—	−30	−30

FIGURE 4.11. Results of calculations and plastic hinge location for stage 4.

A linear elastic analysis of the structure in Figure 4.11 shows a dramatic increase in some of the deflections in the structure. This indicates that collapse of the structure is imminent when the increase in deflections at collapse may amount to tens of thousand times above the results in the previous stage of analysis.

Total collapse load factor, $\alpha_{col} = 0.503 + 0.084 + 0.073 + 0.0072 = 0.6672$ and the total collapse load $= 0.6672 \times 40 = 26.7$ kN.

4.4 Calculation of Design Actions and Deflections

The primary purpose of carrying out plastic analysis is to find the collapse loads of the structure. For a structurally safe structure, the collapse loads must be at or above those at the design load level. However, design actions at ultimate limit states are required only at the design load level for design checks. Design actions, including bending moments, shear forces, and axial forces, are not necessarily equal to those at the collapse load level. The design actions can be obtained by extracting them from results of the elastoplastic analysis. A procedure used to obtain the design actions from an elastoplastic analysis is described. This procedure applies to the calculation of member forces for the strength limit state, as well as deflections for the serviceability limit state.

Because an incremental elastoplastic analysis consists of a series of linear elastic analyses between the formation of plastic hinges, member forces and displacements at the design load level can be found by linear interpolation. The linear interpolation calculation is performed between consecutive plastic hinges, one above and one below the design load level. This is illustrated in Figure 4.12 in which the variation of the bending moments with the applied load (or load factor for multiple loads) of a section is shown.

Figure 4.12 shows that the design load for the structure is P_d. Suppose that the total load for the formation of a plastic hinge in the structure above P_d is P_1, corresponding to a bending moment M_1 in a section, and that below P_d is P_2, corresponding to a bending moment M_2 in the same section. The design moment M_d for the section at design load P_d can be calculated by linear interpolation as

$$M_d = M_2 + \frac{P_d - P_2}{P_1 - P_2}(M_1 - M_2) \tag{4.5}$$

where P_1, M_1, P_2, and M_2 are obtained from elastoplastic analysis. Other forces and displacements can be found in a way similar to Equation (4.5).

Example 4.3 For the portal frame in Example 4.2, find the design moments for all members if the design load for the structure is 24 kN.

Solution. The load factor corresponding to the design load is $\alpha = 24/40 = 0.6$. Hence, the design load lies between stage 2 (total load factor = 0.587) and stage 3 (total load factor = 0.66). The moments for all members at stages 2 and 3 are shown in Table 4.1.

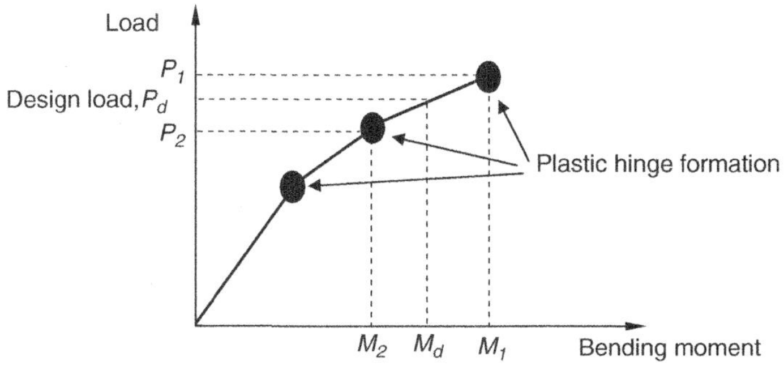

FIGURE 4.12. Interpolation for design actions.

TABLE 4.1
Moments at stages 2 and 3

	Moment (kNm) at stage 2		*Moment (kNm) at stage 3*	
	Joint i	*Joint* j	*Joint* i	*Joint* j
Member 1	19.13	−14.89	28.56	−20
Member 2	−14.89	20	−20	20
Member 3	20	−30	20	−30

TABLE 4.2
Moments at design load level

	Design moment (kNm)	
	Joint i	*Joint* j
Member 1	20.81	−15.80
Member 2	−15.80	20
Member 3	20	−30

The design moments, calculated by linear interpolation using Equation (4.5), for all members at a design load of 24 kN are shown in Table 4.2. An example of the calculation for member 1, joint *i* is illustrated in Figure 4.13.

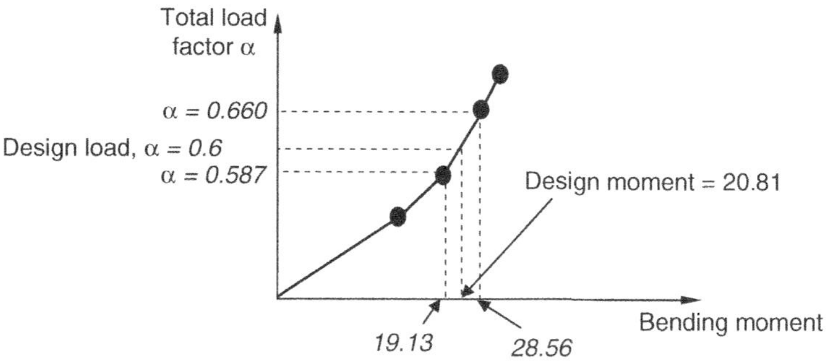

FIGURE 4.13. Interpolation for moments in member 1.

4.5 Effect of Force Interaction on Plastic Collapse

The presence of axial and shear forces in a cross section reduces its plastic moment capacity, which, in turn, reduces the collapse load of the structure. A direct method to take account of the effect of force interaction is to modify the elastoplastic stiffness matrix in accordance with the three cases of yield condition described in Chapter 3 for the yielded section. An indirect method is to assume yielding by pure bending for all elements while the collapse load of the structure is calculated. At the end of the analysis, the reduced plastic moment capacity due to force interaction is calculated for each element and the analysis is repeated. This is called the successive approximation method. Both methods are described here.

4.5.1 Direct Method

This method makes use of the structure stiffness matrix modified to take account of the formation of plastic hinges. The solution of the incremental structure equilibrium equation is based on Equation (3.4) given as

$$\{\Delta D\} = [K_P]^{-1}\alpha\{F\} \tag{4.6}$$

in which $[K_P]$ is the modified structure stiffness matrix. The various forms of the member elastoplastic stiffness matrix $[K_{Pe}]$ lead to different $[K_P]$ being formulated due to the different force interaction formulations and yield conditions. Thus, this method requires special computer programming to create $[K_{Pe}]$ and hence $[K_P]$. In addition, the plastic deformation at the plastic hinge has been condensed into $[K_{Pe}]$ and the extraction of the plastic deformation has to be performed separately. The advantage of using this method is that the force interaction condition is always satisfied at any stage of calculation and the solution for the collapse load is direct.

Calculation of Load Factor

For yield condition based on pure bending, the load factor α for predicting the formation of a plastic hinge at a section is calculated according to Equation (4.3). For yield condition based on force interaction, the calculation of α is more complicated. Its calculation is illustrated in Figure 4.14 in which a yield surface diagram for a section with two dimensionless forces m and β is shown. In addition, the sign of the forces has been considered so that the yield surface is symmetric and consists of four quadrants of hyperplanes.

Suppose that the forces m and β in a section from a linear elastic analysis of the structure under loading $\{F\}$ are represented by the vector OG. Formation of a plastic hinge in the section requires an

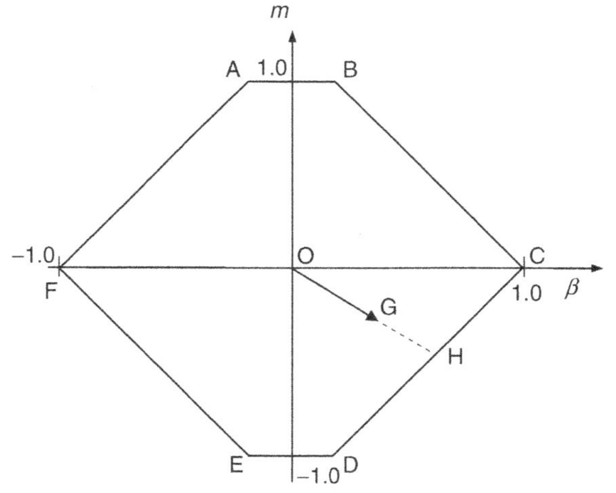

FIGURE 4.14. Yield surface of a section.

increase in the forces by a load factor α such that the vector OG is extended linearly to point H on the hyperplane CD as shown in Figure 4.14. In practice, values of α are calculated for all hyperplanes connecting the points ABCDEF and the one with the smallest positive value is chosen. For the section to stay yielded in subsequent analysis, the force point will move along the yield surface. The following example, used previously in Example 3.2, illustrates the various aspects of this method.

Example 4.4 Determine the collapse load P for the frame ABC shown in Figure 4.15 using the direct method. For both members AB and BC, $E = 2 \times 10^8$ kN/m^2, $A = 0.0104$ m^2, $I = 0.000475$ m^4, $M_p = 515$ kNm, $N_p = 2600$ kN. The members are made of the I section with yield condition given by $m = 1.18(1 - \beta)$ where $m = \dfrac{M}{M_P}$ and $\beta = \dfrac{N}{N_p}$ for $\beta > 0.15$, otherwise $m \leq 1$.

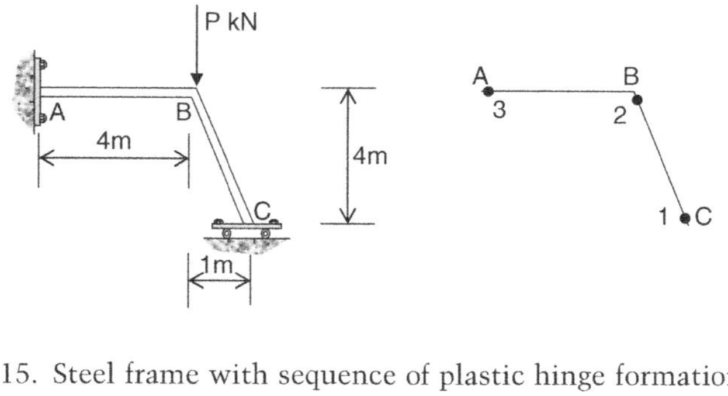

FIGURE 4.15. Steel frame with sequence of plastic hinge formation.

Solution. A load of P = 1 kN is applied to the structure for initial analysis. The aim is to find α for a total collapse load of αP. For this structure, there are three plastic hinges to form before collapse, requiring three stages of analyses. The calculation of α_{cr} is based on geometrical consideration of the hyperplanes shown in Figure 4.16. More discussion on the use of hyperplanes for calculating α_{cr} can be found in Chapter 6. The sequence of plastic hinge formation is shown in Figure 4.15. Results of the calculations for both members AB and BC are shown in Table 4.3.

Movements of the forces on the hyperplane for member BC at different stages of loading are shown in Figure 4.16. For section C, yielding occurs at stage 1 while section B still remains elastic.

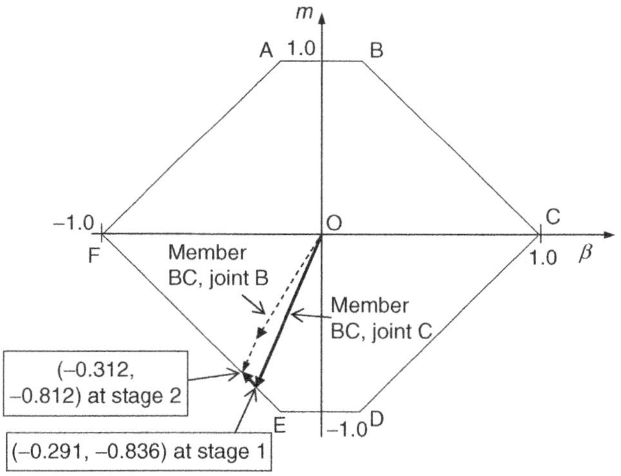

FIGURE 4.16. Movements of forces for member BC.

TABLE 4.3
Results of calculations for Example 4.4

		Member AB			*Member BC*			
	Joint	*Axial force (kN)*	*Bending moment (kNm)*	*m*	*Axial force (kN)*	*Bending moment (kNm)*	*m*	*Critical load factor* (α_{cr})
Stage 1	*i*	0	266.8	0.518	−757.3	−350	−0.680	934.9
	j		350	0.680		−430.7	−0.936	
Stage 2	*i*	0	330.8	0.642	−811.2	−418.1	−0.812	88.5
	j		418.1	0.642		−418.1	−0.812	
Stage 3	*i*		515.0	1	−811.2	−418.1	−0.812	46.0
	j	0	418.1	0.812		−418.1	−0.812	

At stage 2, both sections B and C become yielded and remain at the same point on the hyperplane. The total collapse load for this structure is 1069.4 kN.

4.5.2 Successive Approximation Method

The direct method for elastoplastic analysis requires the use of unique elastoplastic stiffness matrices pertaining to individual yield criteria. Computer software invoking the direct method must be programmed to include these unique formulations. This poses a problem for structural designers using this method as such computer software is not commonly available. As an alternative, a successive approximation method, based on yielding by pure bending in each iterative cycle, can be used to circumvent this problem.

It should be kept in mind that a reduction in the plastic moment capacity in members due to force interaction usually results in a reduction in the plastic collapse load of the structure. When using the successive approximation method, the collapse load factor is calculated on the basis of yielding only by pure bending. Because the total axial forces in the members are not known until the end of the analysis at collapse, the reduced plastic moment capacity as a consequence of axial force or shear force can be calculated only when the analysis is completed. The reduced bending moment capacity for each member can then be calculated and used in a subsequent cycle of analysis. The number of cycles of analysis to be performed depends on the degree of accuracy required for the solution. The method enables the solutions from the analysis cycles to converge to the true collapse load. However, the procedure could be tedious if the structure is complex. An alternative, but conservative, approach is to repeat the cycle of analysis only once. The result would underestimate the collapse load and err on the safe side for design. The procedure for this method is demonstrated in the following example using the structure shown in Example 4.4.

Example 4.5 Determine the collapse load P for the frame ABC shown in Figure 4.15 using the successive approximation method. For both members AB and BC, $E = 2 \times 10^8\ \text{kN/m}^2$, $A = 0.0104\ \text{m}^2$, $I = 0.000475\ \text{m}^4$, $M_p = 515\ \text{kNm}$, $N_p = 2600\ \text{kN}$. The members are made of I section with the yield condition given by $m = 1.18(1-\beta)$ where $m = \dfrac{M}{M_P}$ and $\beta = \dfrac{N}{N_P}$ for $\beta > 0.15$, otherwise $m \leq 1$.

Solution. For each cycle of analysis, the critical load factor is calculated using the pure bending yield criterion [Equation (4.3)]. Results of the first cycle of analysis are given in Table 4.4. At the end of iteration cycle 1, the collapse load of the structure is 1287.5 kN and the total axial force in BC is 999.3 kN.

TABLE 4.4
Results of calculations for Example 4.5: Iteration cycle 1

		Member AB		*Member BC*		
	Joint	*Axial force (kN)*	*Bending moment (kNm)*	*Axial force (kN)*	*Bending moment (kNm)*	*Critical load factor (α_{cr})*
Stage 1	*i*	0	319.1	−905.7	−418.5	1117.9
	j		418.5		−515.0	
Stage 2	*i*	0	400.9	−999.3	−515.0	141.1
	j		515.0		−515.0	
Stage 3	*i*	0	515.0	−999.3	−515.0	28.5
	j		515.0		−515.0	

The reduced bending moment capacity of member BC is calculated and found to be 374 kNm, which is used in the next iteration cycle of analysis while the bending moment capacity of member AB remains at 515 kNm. Results of iteration cycle 2 are shown in Table 4.5.

At the end of iteration cycle 2, the total collapse load is 970.3 kN and the total axial force in BC is 725.7 kN. The reduced bending moment capacity of member BC is calculated and found to be 438 kNm, which is used in the next iteration cycle of analysis.

The aforementioned iteration cycle can be repeated as many times as the accuracy of solution requires. For this example, the collapse load is 1071.5 kN after seven iteration cycles. The iteration trend is shown in Figure 4.17, where the collapse load is shown to converge on the exact solution of 1069.4 kN.

TABLE 4.5
Results of calculations for Example 4.5

		Member AB		*Member BC*		
	Joint	*Axial force (kN)*	*Bending moment (kNm)*	*Axial force (kN)*	*Bending moment (kNm)*	*Critical load factor (α_{cr})*
Stage 1	*i*	0	231.7	−657.7	−303.9	811.9
	j		303.9		−374.0	
Stage 2	*i*	0	291.1	−725.7	−374.0	102.4
	j		374.0		−374.0	
Stage 3	*i*	0	515.0	−725.7	−374.0	56.0
	j		374.0		−374.0	

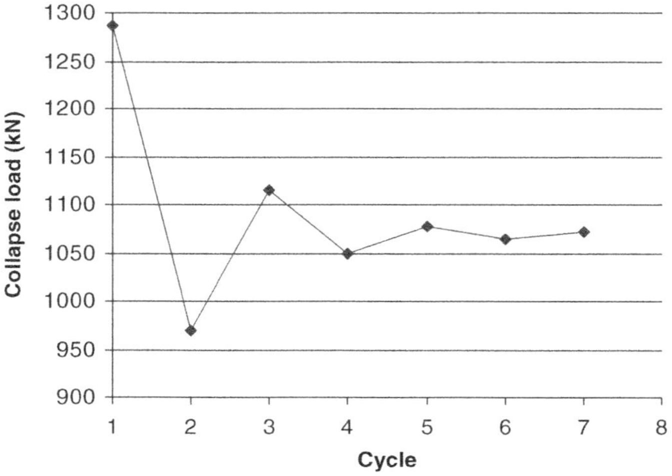

FIGURE 4.17. Convergence of solution using successive approximation method.

4.6 Plastic Hinge Unloading

In incremental elastoplastic analysis, it is usually assumed that the plastic hinges, once formed, remain yielded throughout the analysis. This is true for many structures. However, for some structures, plastic hinges may unload and become elastic again due to the formation of plastic hinges elsewhere in the structure. The detection of plastic hinge unloading can be performed by calculating the plastic multiplier λ for each plastic hinge as described in Chapter 3. A plastic hinge undergoes unloading when the value of λ is negative. At any stage of analysis when unloading is detected, the section corresponding to the unloading plastic hinge becomes elastic and that stage of analysis is repeated with the section modified as an elastic one. The following structure demonstrates plastic hinge unloading in the process of increasing loads.

Example 4.6 Determine the collapse load factor P for the single-story, double-bay frame shown in Figure 4.18 using the direct method. For all members, $E = 2.1 \times 10^8$ kN/m^2 and yielding is based only on pure bending.

For members AB and HG, $A = 0.003$ m^2, $I = 0.00001$ m^4, $M_p = 45$ kNm.
For members BC, CD, DF, and FG, $A = 0.004$ m^2, $I = 0.00004$ m^4, $M_p = 78$ kNm.
For member ED, $A = 0.002$ m^2, $I = 0.000005$ m^4, $M_p = 17.5$ kNm.

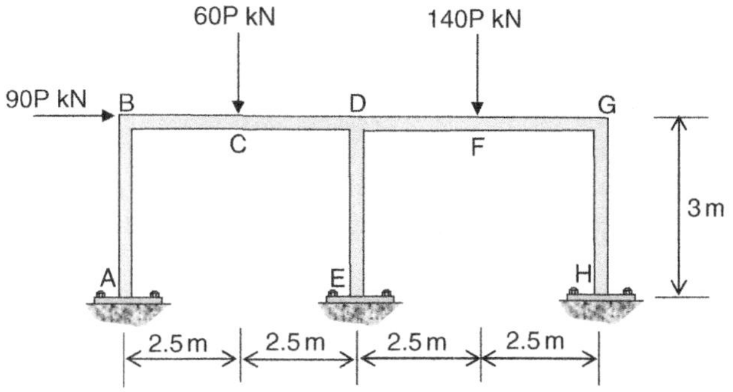

FIGURE 4.18. Example 4.6.

Solution. A curve for the variation of P with the horizontal deflection at B and the sequence of plastic hinge formation for the structure is shown in Figure 4.19. The analysis shows that hinge 4 undergoes unloading after hinge 5 in member FG occuring at a total load factor P = 0.735. Also, hinge 2 at E undergoes unloading after hinge 6 in member CD occuring at a total load factor P = 0.747. The section at E becomes a plastic hinge again when hinge 8 occurs at a total load factor P = 0.752. At collapse, the total load factor P = 0.769.

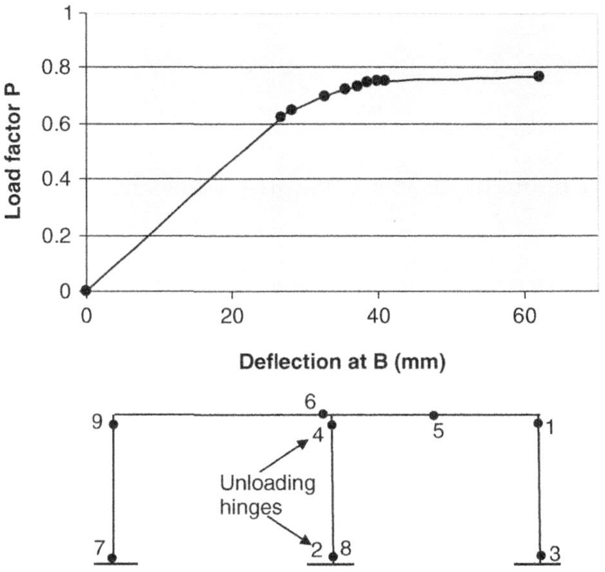

FIGURE 4.19. Load–deflection curve and hinge formation sequence for Example 4.6.

4.7 Distributed Loads in Elastoplastic Analysis

The plastic analysis methods described so far have focused on structures subjected to point loads only. For structures subjected to point loads, plastic hinges occur at sections either at joints or where the point loads act. A member with internal point loads can be discretized into shorter elements connected at joints where the internal point loads are acting.

For members subjected to distributed load, plastic hinges may occur within the length of the load along the member. The usual way of dealing with distributed load in plastic analysis is to simulate the action of the distributed load by equivalent point loads. The member subjected to the distributed load is then discretized into shorter elements according to the number of equivalent point loads generated. Although this method is usually satisfactory, the difficulty of determining an adequate number of equivalent point loads is always present, and the accuracy of results is not known. In the following paragraph, an iterative procedure to locate the plastic hinge for members under distributed loads is presented. The method shows convergence both to the correct locations of the plastic hinges in members and to the correct plastic collapse load.

A member of length L subjected to a distributed load linearly varying from w_1 at one end to w_2 at the other is shown in Figure 4.20.

Assume that the shear force and bending moment at end i of the member are S_i and M_i, respectively, obtained at the completion of an elastoplastic analysis. The bending moment M_x at a distance x from end i is given by

$$M_x = S_i x - \frac{w_1 x^2}{2} - \frac{(w_2 - w_1)x^3}{6L} - M_i \tag{4.7}$$

The bending moment M_x is a maximum when the shear force is zero. That is,

$$S_x = \frac{\partial M_x}{\partial x} = S_i - w_1 x - \frac{(w_2 - w_1)x^2}{2L} = 0 \tag{4.8}$$

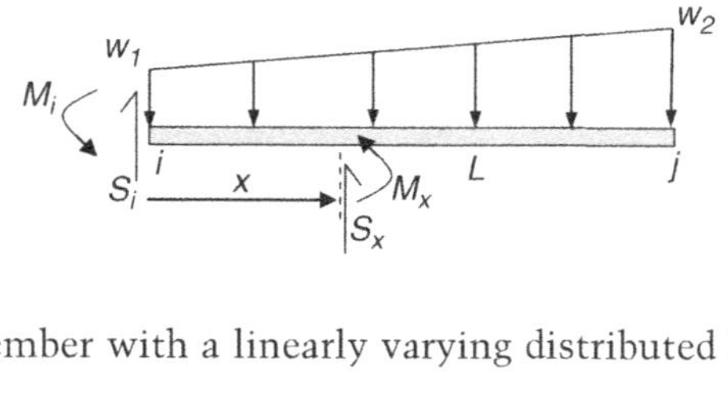

FIGURE 4.20. Member with a linearly varying distributed load.

By solving Equation (4.8) for x, the location for the occurrence of a possible plastic hinge is identified. The member can be discretized into two with the joint located at the point where maximum M_x occurs and the collapse load can be reevaluated. This procedure can be repeated until both the location of the plastic hinge and the collapse load of the structure converge with satisfactory accuracy.

Example 4.7 Determine the plastic collapse load factor α and the location of the internal plastic hinge at a distance x from A for the fixed-end beam ABC with a UDL of 10 kN/m along AB as shown in Figure 4.21. Plastic moment capacity of the member, $M_p = 100$ kNm.

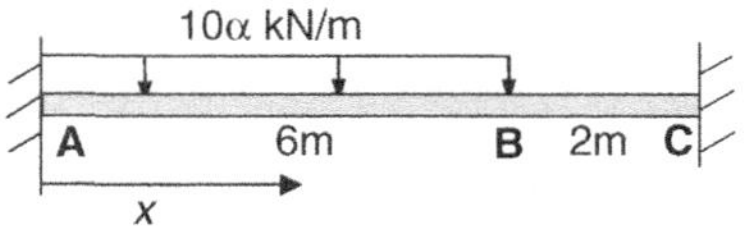

FIGURE 4.21. Example 4.7.

Solution. The beam is discretized into three elements so that a joint can be arbitrarily assigned within the length AB. In the first instance, a joint D is assigned at a distance of 3 m from A and the resulting plastic hinge formation sequence is shown in Figure 4.22a. Elastoplastic analysis cycle 1 gives a total collapse load factor $\alpha = 2.963$. Using Equation (4.8), the location of the maximum bending moment is found to be 3.75 m from A in element DB and the corresponding bending moment is 108.3 kNm from Equation (4.7). Obviously, the bending moment is in violation of the yield condition.

A second elastoplastic analysis is then carried out with a revised location for the joint D at 3.75 m from A. In this case, the total collapse load factor is found to be 2.844 and the location of the maximum bending moment remains unchanged at D as shown in Figure 4.22b. It should be noted that the moving joint D is where the last plastic hinge occurs. Therefore, in this example, only one adjustment of the plastic hinge location is required to give the correct collapse load factor using only one iteration of analysis. A summary of the results is given in Table 4.6.

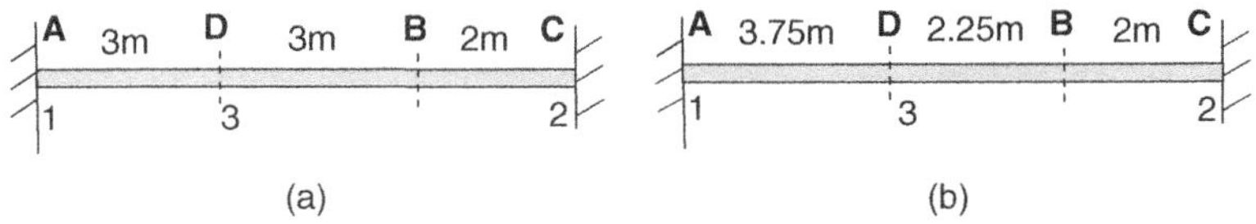

FIGURE 4.22. Member discretization and hinge formation sequence.

TABLE 4.6
Results of iterative analysis

Analysis cycle	*Hinge formation sequence*	*Total load factor α*	*Plastic hinge location* x *from A (m)*
1	1	1.975	3.0
	2	2.319	
	3	2.963	
2	1	1.975	3.75
	2	2.319	
	3	2.844	

It may be useful to compare the results just given with those obtained from the equivalent point loads method. When using this method, a number of alternatives may convert the distributed load into point loads. Some of these load conversions and the corresponding plastic hinge formation sequence are shown in Figure 4.23. Results are shown in Table 4.7. It can be seen that the structure shown in Figure 4.23b gives results comparable to those obtained using the more accurate iterative method. When using the equivalent point loads method, the accuracy of the results depends on both the number of discretized elements and the way the equivalent point loads are calculated. In general, the accuracy increases with the number of discretized elements used. Note that the order of plastic hinge formation is not the same for all the structures shown in Figure 4.23.

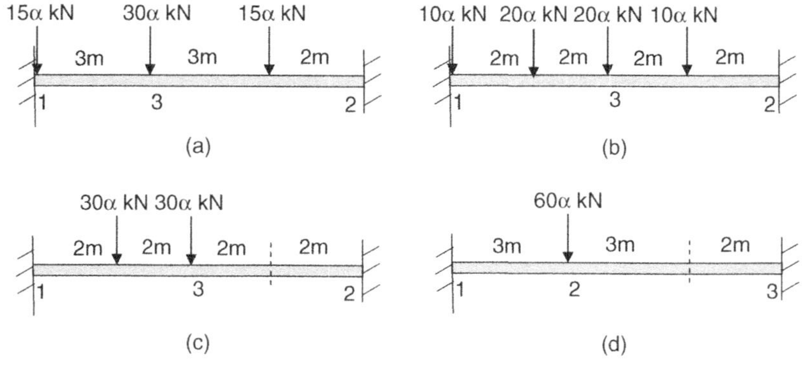

FIGURE 4.23. Equivalent point loads and hinge formation sequence.

TABLE 4.7
Results from equivalent point load method

Structure	*Hinge formation sequence*	*Total load factor α*
Figure 4.23a	1	2.452
	2	2.570
	3	2.963
Figure 4.23b	1	2.162
	2	2.424
	3	2.857
Figure 4.23c	1	1.568
	2	2.051
	3	2.222
Figure 4.23d	1	1.422
	2	1.721
	3	1.778

Example 4.8 Determine the collapse load factor α of the portal frame shown in Figure 4.24 Properties of the members are

Columns AB, ED, HG: $A = 0.02\ m^2$, $I = 0.01\ m^4$, $M_p = 20$ kNm, $N_p = 80$ kN.
Beam BD: $A = 0.04\ m^2$, $I = 0.015\ m^4$, $M_p = 50$ kNm, $N_p = 120$ kN.
Beam DG: $A = 0.07\ m^2$, $I = 0.02\ m^4$, $M_p = 80$ kNm, $N_p = 200$ kN.

The members are made of the I section with the yield condition given by $m = 1.18(1-\beta)$ where $m = \dfrac{M}{M_p}$ and $\beta = \dfrac{N}{N_p}$ for $\beta > 0.15$, otherwise $m \leq 1$.

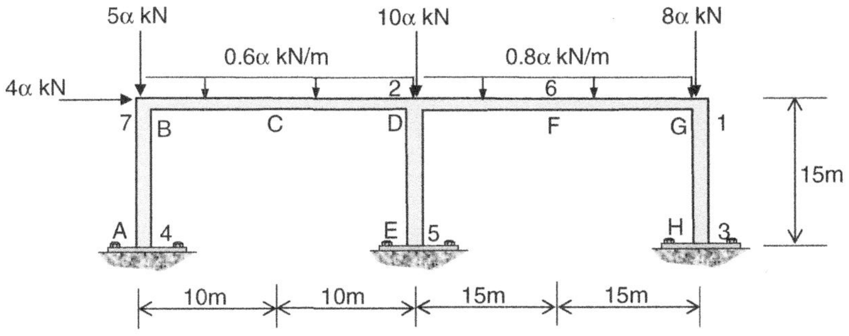

FIGURE 4.24. Example 4.8 with plastic hinge formation sequence.

Solution. The beams are initially assigned internal nodes at midspan at C and F in BD and DG, respectively. The sequence of plastic hinge formation, with a total of seven plastic hinges, is also shown in Figure 4.24. The first attempt gives a collapse load factor $\alpha = 1.2935$, and the maximum moment location for DG is found to be 16.296 m from D. By relocating the node F to this point at 16.296 m from D and performing a second analysis, collapse load factor α is found to be 1.2868. Subsequent analysis shows that the relocated node F remains unchanged. Note that for all members made of the same material, the collapse load is independent of the Young's modulus, *E*.

Problems

4.1. Determine the collapse load factor α for the propped cantilever beam ABCD shown in Figure P4.1. $M_p = 80$ kNm; E, I, and A are constants.

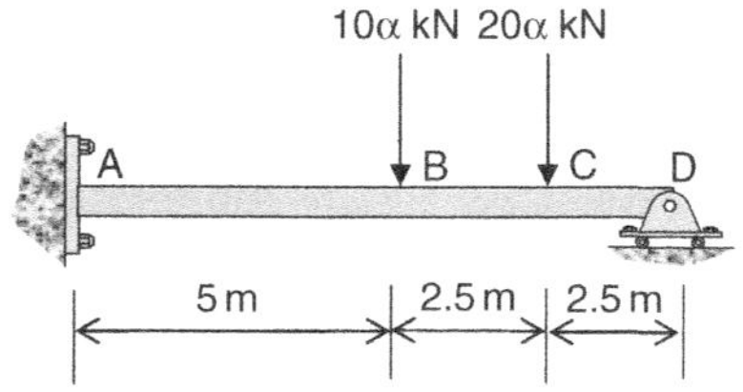

FIGURE P4.1. Problem 4.1.

4.2. A steel warehouse is subjected to a horizontal design load of 150 kN acting at B and a vertical design load of 110 kN acting at C shown in Figure P4.2. In order to find the collapse strength of the structure, the loads are assumed to increase proportionally by a common factor α. An initial sizing for the members requires

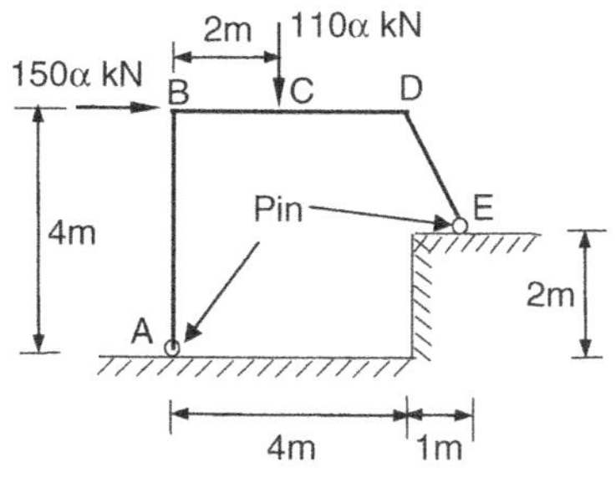

FIGURE P4.2. Problem 4.2.

a steel section of plastic moment capacity equal to 140 kNm to be used for all members. An elastoplastic analysis is then carried out for the proportional loads applied at $\alpha = 1$. Results of the analysis for bending moments and horizontal deflection at B are shown in the following tables.

Calculate the total bending moments for all members, the total horizontal deflection at B at collapse, and the collapse load factor α_c of the structure. Also, calculate the bending moment and the horizontal deflection, both at B, at design load level (i.e., at total $\alpha = 1.0$). Assume pure bending for the yield condition.

Analysis Stage No: 1 Critical Load Factor, $\alpha_{cr} =$

Member	*Joint*	*Moment* M_o *(kNm)*	*Residual plastic moment* M_P-M_i *(kNm)*	*Load factor* $\alpha = \frac{M_p - M_i}{M_o}$	*Cumulative moment* $M_{i+1} = M_i + \alpha_{cr} M_o$ *(kNm)*	*Horizontal deflection* d_o *(mm)*	*Cumulative horizontal deflection* $q_{i+1} = q_i + \alpha_{cr} d_o$ *(mm)*
AB	A	0				30	
	B	−46.2					
BC	B	−46.2					
	C	−51.1					
CD	C	−51.1					
	D	166.6					
DE	D	166.6					
	E	0					

Analysis Stage No: 2 Critical Load Factor, $\alpha_{cr} =$

Member	*Joint*	*Moment* M_o *(kNm)*	*Residual plastic moment* $M_P - M_i$ *(kNm)*	*Load factor* $\alpha = \frac{M_p - M_i}{M_o}$	*Cumulative moment* $M_{i+1} = M_i + \alpha_{cr} M_o$ *(kNm)*	*Horizontal deflection* d_o *(mm)*	*Cumulative horizontal deflection* $q_{i+1} = q_i + \alpha_{cr} d_o$ (mm)
AB	A	0				234	
	B	−323.9					
BC	B	−323.9					
	C	−273.3					
CD	C	−273.3					
	D	0					
DE	D	0					
	E	0					

Analysis stage 3 shows that the structure has collapsed.

4.3. A steel lecture theatre is subjected to design loads shown in Figure P4.3. The plastic method has been adopted for the design of the structure. In order to find the failure load of the structure, the loads are assumed to increase proportionally by common factor α.

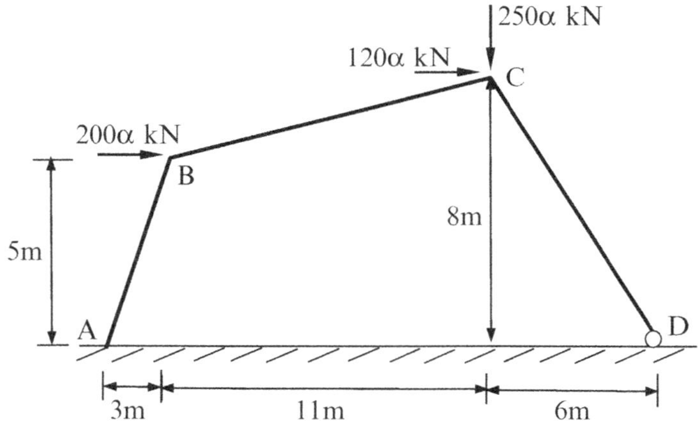

FIGURE P4.3. Problem 4.3.

An initial sizing for the members requires a steel section with $M_p = 304$ kNm and $N_p = 1810$ kN to be used for all members. An elastoplastic analysis is then carried out for loads applied at the design level (i.e., $\alpha = 1$). Results of the analysis for bending moments M_o and axial forces N_o are shown:

Analysis Stage No: 1 Critical Load Factor, $\alpha_{cr} =$

Member	*Joint*	*Moment* M_o *(kNm)*	*Residual plastic moment* $M_P - M_i$ *(kNm)*	*Load factor* $\alpha = \frac{M_p - M_i}{M_o}$	*Cumulative moment* $M_{i+1} = M_i + \alpha_{cr} M_o$ *(kNm)*	*Axial force (kN)* N_o	*Cumulative axial force (kN)* $N_{i+1} = N_i + \alpha_{cr} N_o$
AB	A	338.3				63.7	
	B	−211.9					
BC	B	−211.9				−81.7	
	C	114.0					
CD	C	114.0				−328.7	
	D	0					

Note: Negative (−) axial force = compression.

Analysis Stage No: 2 Critical Load Factor, α_{cr} =

Member	*Joint*	*Moment* M_o *(kNm)*	*Residual plastic moment* $M_P - M_i$ *(kNm)*	*Load factor* $\alpha = \frac{M_p - M_i}{M_o}$	*Cumulative moment* $M_{i+1} = M_i + \alpha_{cr} M_o$ *(kNm)*	*Axial force (kN)* N_o	*Cumulative axial force (kN)* $N_{i+1} = N_i + \alpha_{cr} N_o$
AB	A	0				63.8	
	B	−359.7					
BC	B	−359.7				−104.2	
	C	236.1					
CD	C	236.1				−359.0	
	D	0					

Analysis Stage No: 3 Critical Load Factor, α_{cr} =

Member	*Joint*	*Moment* M_o *(kNm)*	*Residual plastic moment* $M_P - M_i$ *(kNm)*	*Load factor* $\alpha = \frac{M_p - M_i}{M_o}$	*Cumulative moment* $M_{i+1} = M_i + \alpha_{cr} M_o$ *(kNm)*	*Axial force (kN)* N_o	*Cumulative axial force (kN)* $N_{i+1} = N_i + \alpha_{cr} N_o$
AB	A	0				26.8	
	B	0					
BC	B	0				−173.6	
	C	811.8					
CD	C	811.6				−402.1	
	D	0					

Analysis stage 4 shows that the structure has collapsed.

a. Calculate the total bending moments and the total axial forces at collapse for all members.
b. By calculating the critical load factor α_{cr} for each stage of analysis, determine the collapse load factor α_{col} of the structure.
c. Calculate the bending moment (M^*) and the axial force (N^*) for member CD at design load level (i.e., at total $\alpha = 1.0$).
d. Check whether member CD would satisfy the following combined bending-axial actions for design:

$$M^* \leq 1.18 M_p \left(1 - \frac{N^*}{N_p}\right) \leq M_p$$

4.4. Determine the collapse load factor α for the portal frame shown in Figure P4.4. For all members, $E = 2 \times 10^7$ kN/m^2, $A = 0.0104$ m^2, $I = 0.000475$ m^4, $M_p = 515$ kNm, $N_p = 2600$ kN. The members are made of rectangular sections with yield condition approximated by $m = 1.43(1 - \beta)$ where $m = \dfrac{M}{M_P}$ and $\beta = \dfrac{N}{N_P}$ for $\beta > 0.3$, otherwise $m \leq 1$.

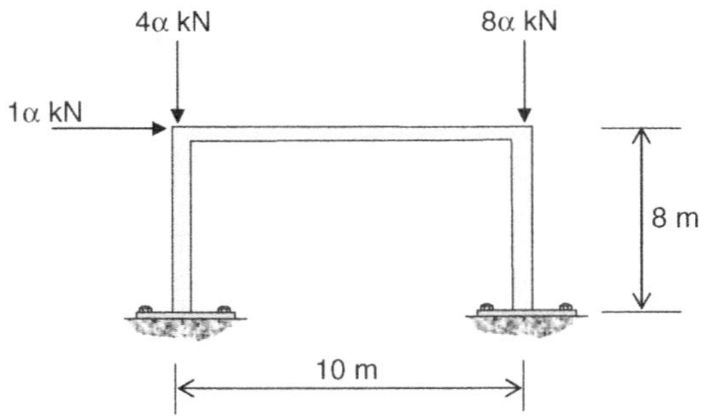

FIGURE P4.4. Problem 4.4.

4.5. Determine the collapse load factor α for the pin-based portal frame shown in Figure P4.5. For all members, E = 250000, A = 0.04, I = 0.15, $M_p = 200$, $N_p = 700$, all in consistent units. The members are made of I sections with yield condition given by $m = 1.18(1 - \beta)$ where $m = \dfrac{M}{M_P}$ and $\beta = \dfrac{N}{N_P}$ for $\beta > 0.15$, otherwise $m \leq 1$.

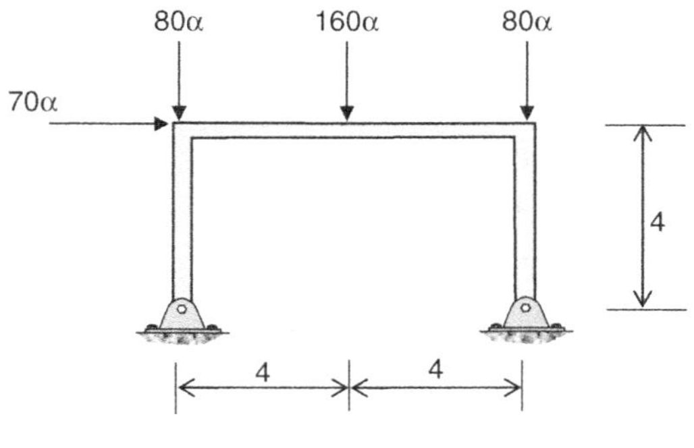

FIGURE P4.5. Pin-based portal frame.

4.6. Determine the collapse load factor α for the gable frame shown in Figure P4.6. For all members, $E = 2 \times 10^8$ kN/m^2. For members AB and ED, A = 0.0179 m^2, I = 0.00136 m^4, M_p = 1140 kNm, N_p = 4475 kN. For members BC and CD, A = 0.0188 m^2, I = 0.00169 m^4, M_p = 1290 kNm, N_p = 4700 kN. The members are made of I sections with yield condition given by $m = 1.18(1-\beta)$ where $m = \dfrac{M}{M_P}$ and $\beta = \dfrac{N}{N_P}$ for $\beta > 0.15$, otherwise $m \leq 1$.

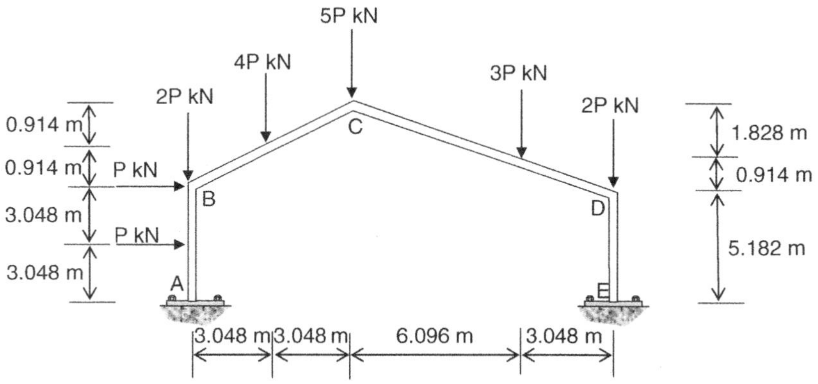

FIGURE P4.6. Gable frame.

4.7. Determine the load factor α and the horizontal deflection at C at collapse for the two-story, rigid-jointed frame shown in Figure P4.7. For all members, $E = 2.1 \times 10^8$ kN/m^2. For members AB, BC, DE, and EF, $A = 0.145525$ m^2, $I = 0.024604$ m^4, $M_p = 11290$ kNm,

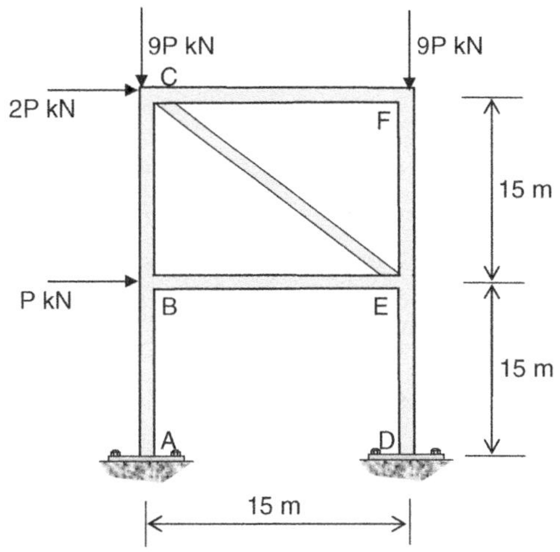

FIGURE P4.7. Two-story rigid-jointed frame.

$N_p = 29100$ kN. For members CF, CE, and BE, A $= 0.022996\ m^2$, I $= 0.001008\ m^4$, $M_p = 893$ kNm, $N_p = 4600$ kN. The members are made of tubular sections with yield condition approximated by $m = 1.25(1 - \beta)$ where $m = \dfrac{M}{M_P}$ and $\beta = \dfrac{N}{N_P}$ for $\beta > 0.2$, otherwise $m \leq 1$.

4.8. Determine the collapse load factor α for the propped cantilever beam ABC subjected to UDL of 10α kN/m along BC shown in Figure P4.8. Locate the plastic hinges at collapse. $M_p = 80$ kNm; E, I, and A are constants.

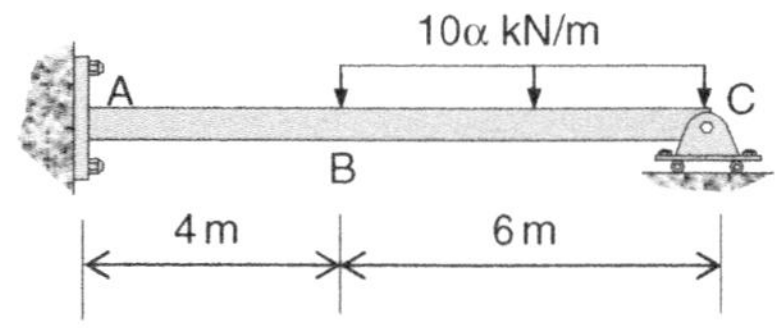

FIGURE P4.8. Propped cantilever under UDL.

4.9. A frame ABCD shown in Figure P4.9 is analyzed with different pinned conditions at the joints, and results of the bending moments are shown in Figure P4.10. If the plastic moment capacity of the members of the frame is 169 kNm, determine the sequence of plastic hinge formation and hence the load factor α at collapse.

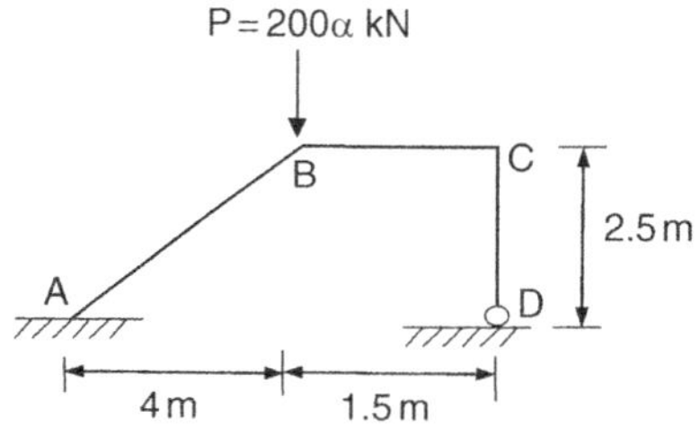

FIGURE P4.9. Problem 4.9.

Case	Structure	Joint A	Joint B	Joint C	Joint D
A	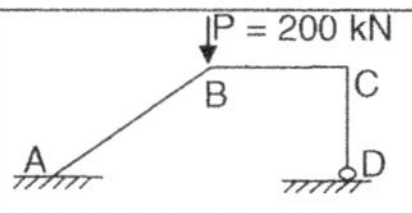	72.3	−98.5	100.0	0
B	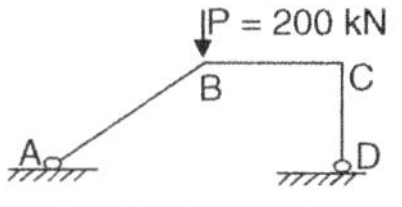	0	−92.2	126.0	0
C	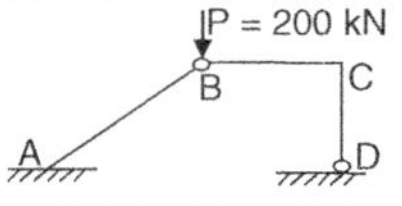	51.9	0	204.0	0
D	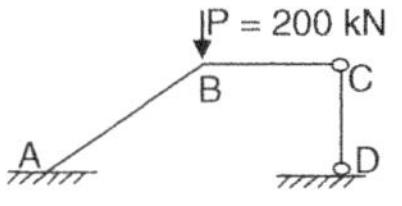	138.4	−180.4	0	0
E	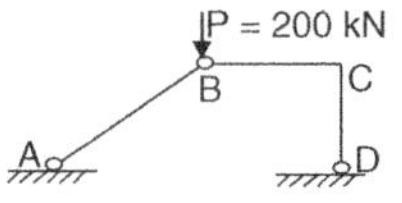	0	0	218.2	0
F	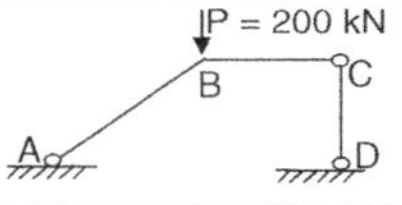	0	−218.2	0	0
G	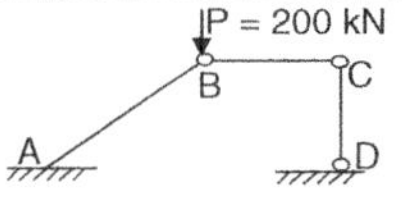	800	0	0	0
H	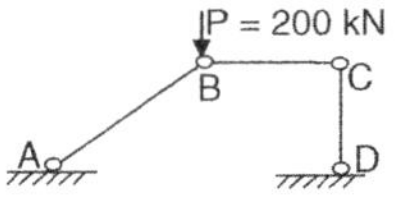	Collapse			

FIGURE P4.10. Results of analyses of different joint cases.

Bibliography

1. Wong, M. B. (1991). Direct analytical methods for plastic design. *Civil Engineering Transaction, IEAust,* **CE33**(2), pp. 81–86.

CHAPTER 5

Manual Methods of Plastic Analysis

5.1 Introduction

In contrast to incremental elastoplastic analysis, classical rigid plastic analysis has been used for plastic design over the past decades, and textbooks on this topic are abundant.[1–3] Rigid plastic analysis makes use of the assumption that the elastic deformation is so small that it can be ignored. Therefore, in using this method of analysis, the material behaves as if the structure does not deform until it collapses plastically. This behavior is depicted in the stress–strain diagram shown in Figure 5.1.

Although classical rigid plastic analysis has many restrictions in its use, its simplicity still has certain merits for the plastic design of simple beams and frames. However, its use is applicable mainly for manual calculations as it requires substantial personal judgment to, for instance, locate the plastic hinges in the structure. This sometimes proves to be difficult for inexperienced users. This chapter describes the classical theorems of plasticity. The applications of these theorems to plastic analysis are demonstrated by the use of mechanism and statical methods, both of which are suitable for manual calculations of simple structures. Emphasis is placed on the use of the mechanism method in which rigid plastic behavior for steel material is assumed.

5.2 Theorems of Plasticity

There are three basic theorems of plasticity from which manual methods for collapse load calculations can be developed. Although attempts have been made to generalize these methods by computers,[4]

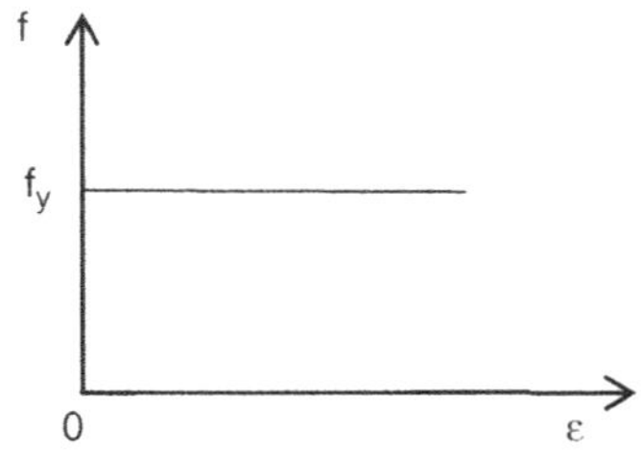

FIGURE 5.1. Rigid plastic behavior.

the calculations based on these methods are still largely performed manually. The basic theorems of plasticity are kinematic, static, and uniqueness, which are outlined next.

5.2.1 Kinematic Theorem (Upper Bound Theorem)

This theorem states that the collapse load or load factor obtained for a structure that satisfies all the conditions of yield and collapse mechanism is either greater than or equal to the true collapse load. The true collapse load can be found by choosing the *smallest* value of collapse loads obtained from all possible cases of collapse mechanisms for the structure. The method derived from this theorem is based on the balance of external work and internal work for a particular collapse mechanism. It is usually referred to as the mechanism method.

5.2.2 Static Theorem (Lower Bound Theorem)

This theorem states that the collapse load obtained for a structure that satisfies all the conditions of static equilibrium and yield is either less than or equal to the true collapse load. In other words, the collapse load, calculated from a collapse mode other than the true one, can be described as conservative when the structure satisfies these conditions. The true collapse load can be found by choosing the *largest* value of the collapse loads obtained from all cases of possible yield conditions in the structure. The yield conditions assumed in the structure do not necessarily lead to a collapse mechanism for the structure. The use of this theorem for calculating the collapse load of an indeterminate structure usually considers static equilibrium through a flexibility approach to produce free and reactant bending moment diagrams. It is usually referred to as the statical method.

5.2.3 Uniqueness Theorem

It is quite clear that if a structure satisfies the conditions of both static and kinematic theorems, the collapse load obtained must be true and unique. Therefore, the uniqueness theorem states that a true collapse load is obtained when the structure is under a distribution of bending moments that are in static equilibrium with the applied forces and no plastic moment capacity is exceeded at any cross section when a collapse mechanism is formed. In other words, a unique collapse load is obtained when the three conditions of *static equilibrium, yield,* and *collapse mechanism* are met.

It should be noted that an incremental elastoplastic analysis such as that described in Chapter 4 satisfies all three of these conditions: (1) static equilibrium—elastic analysis is based on solving a set of equilibrium equations contained in matrices; (2) yield—the moment capacity for every section is checked and a plastic hinge is inserted if plastic moment is reached in any section; insertion of a plastic hinge in the analysis ensures that the moment capacity is not exceeded; and (3) mechanism—the formation of a collapse mechanism is checked by (a) determining whether the determinant of the stiffness matrix is zero; a zero value leads to an error message if a computer is used for analysis; and (b) excessive deflections if an exact zero stiffness cannot be detected. Hence, the collapse load obtained from an elastoplastic analysis is, in general, unique.

5.3 Mechanism Method

This method requires that all possible collapse mechanisms are identified and that the virtual work equation for each mechanism is established. The collapse load P_w (or collapse load factor α_c if a set of loads are applied) is the minimum of the solutions of all possible collapse mechanisms for the structure. In establishing the virtual work equation, the total internal work as sum of the products of the plastic moment, M_p, and the corresponding plastic rotation, θ, at all plastic hinge locations j (j = A, B, ..., etc.) must be equal to the total external work. The total external work is expressed as the sum of the products of the externally applied load, $\alpha_c P$, and the corresponding distance, δ, it displaces for all loads i (i = 1, 2, ..., etc.). Mathematically,

$$\sum_j (M_p\theta)_j = \alpha_c \sum_i (P\delta)_i \tag{5.1}$$

For Equation (5.1), a relationship between θ and δ can be established so that α_c is evaluated independently of these two terms.

Example 5.1 A fixed-end beam, of length L and plastic moment capacity M_p, is subject to a point load P as shown in Figure 5.2a. Determine the collapse load $P = P_w$.

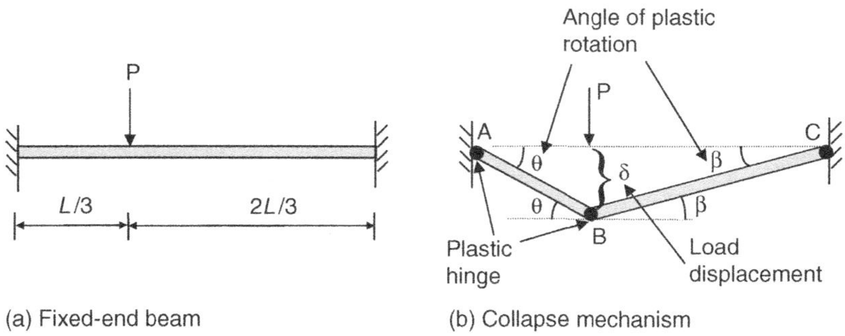

FIGURE 5.2. Plastic analysis of a fixed-end beam.

Solution. The first step is to guess the possible collapse mechanism. In this case, it is obvious that the collapse mechanism is induced by the formation of three plastic hinges at A, B, and C shown in Figure 5.2b.

From Figure 5.2b, the load displacement can be related approximately to the angles of plastic rotation by

$$\delta = \frac{L}{3}\theta = \frac{2L}{3}\beta \qquad (5.2)$$

Hence, $\theta = 2\beta$. At B, the total plastic rotation is $(\theta + \beta)$. The internal work, W_i, of the system due to the plastic rotations at A, B, and C is

$$W_i = M_P\theta + M_P(\theta + \beta) + M_P\beta = M_P 6\beta \qquad (5.3)$$

The external work, W_e, is

$$W_e = P_w\delta = P_w\frac{2L}{3}\beta \qquad (5.4)$$

The virtual work equation is $W_e = W_i$, which gives $P_w = \dfrac{9M_P}{L}$.

5.4 Statical Method

This method uses the flexibility (or force) method of analysis, which requires construction of the final bending moment diagram by superposing a free (from a determinate structure) bending moment diagram upon the reactant (from a redundant structure) bending moment diagrams. As for the mechanism method, only a guess of the correct collapse mechanism will lead to the correct estimate of the collapse load when using the statical method.

Example 5.2 A cantilever beam ABCD is subject to two point loads of 30α and 50α kN shown in Figure 5.3. The plastic moment capacity of the beam is M_p. Calculate the collapse load factor $\alpha = \alpha_c$.

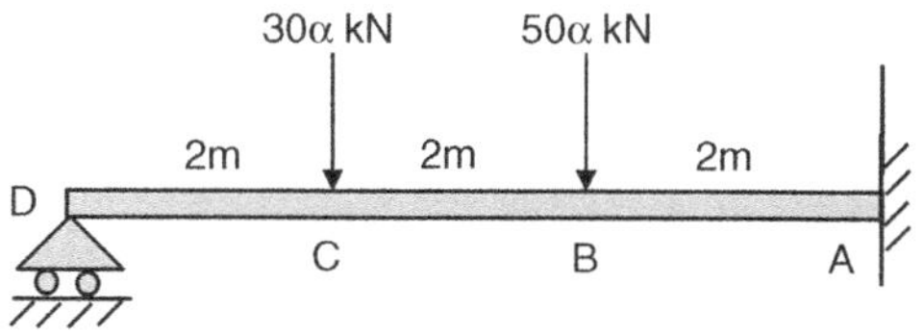

FIGURE 5.3. Plastic analysis of cantilever beam using the statical method.

Solution. This is an indeterminate structure with 1 degree of indeterminacy. If the bending moment M at the fixed support A is made redundant, then the free and reactant bending moment diagrams can be constructed as shown in Figure 5.4.

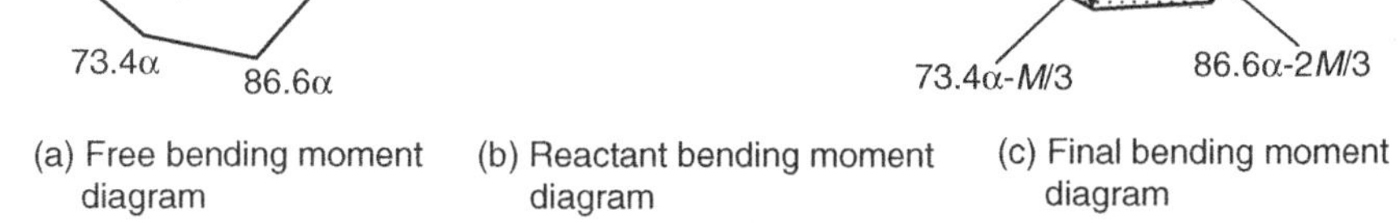

FIGURE 5.4. Free and reactant bending moment diagrams.

This structure requires the formation of two plastic hinges to induce a collapse mechanism. There are two possibilities: (a) plastic hinges at A and B or (b) plastic hinges at A and C. It is unlikely to have a collapse mechanism with plastic hinges at B and C because of the larger moment at A.

Case a: plastic hinges at A and B.

$$M = M_p \qquad \text{and} \qquad M_p = 86.6\alpha - \frac{2}{3}M,$$

therefore $$\alpha = \frac{M_p}{52}.$$

Check: Bending moment at

$$C = 73.4\alpha - \frac{1}{3}M_p = 73.4 \times \frac{M_p}{52} - \frac{1}{3}M_p = 1.08M_p.$$

This case is not valid because the yield condition is violated at C.

Case b: plastic hinges at A and C.

$$M = M_p \qquad \text{and} \qquad M_p = 73.4\alpha - \frac{1}{3}M,$$

therefore $$\alpha = \frac{M_p}{55}.$$

Check: Bending moment at

$$B = 86.6\alpha - \frac{2}{3}M_p = 73.4 \times \frac{M_p}{55} - \frac{2}{3}M_p = 0.67M_p.$$

The yield condition has not been violated at B. Hence, case b is critical and $\alpha = \alpha_c = \frac{M_p}{55}$.

It should be noted that, according to the static theorem, the true collapse load from the statical method is the largest of all possible solutions only if the yield or equilibrium condition is not violated. The static theorem does not apply to this example as the yield condition in case a is violated.

5.5 Uniformly Distributed Loads (UDL)

When using the mechanism method, the main difficulty in dealing with a distributed load is to calculate the external work as it normally requires integration for its evaluation. However, some convenient concepts can be developed to circumvent this difficulty. The following example demonstrates the treatment of uniformly distributed loads using both statical and mechanism methods.

Example 5.3 A fixed-end beam shown in Figure 5.5 is subjected to a uniformly distributed load of w (load/length). Determine the collapse load $w = w_c$.

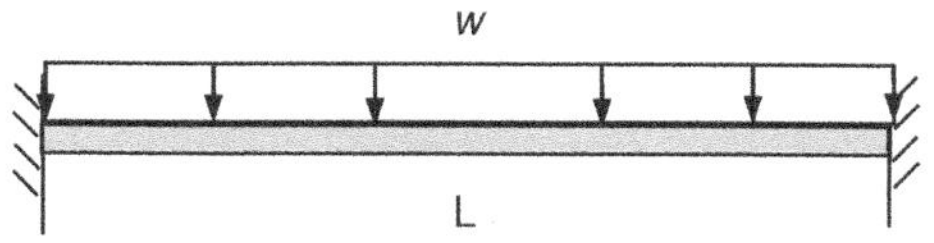

FIGURE 5.5. Fixed-end beam with UDL.

Solution

By Statical Method

Fixed-end moments, which are equal because of symmetry, are chosen as the redundant forces. The free and reactant bending moment diagrams are shown in Figure 5.6.

For this problem, three plastic hinges, two at the ends and one at midspan of the beam, are required to induce a collapse mechanism. From Figure 5.6,

$M = M_P$ at the supports and $\frac{wL^2}{8} - M = M_p$ at midspan.

therefore $w = w_c = \frac{16M_p}{L^2}$

By Mechanism Method

The collapse mechanism is shown in Figure 5.7. For the two straight segments of the beam (half of the beam in this case), total external work $= 2 \times \int_0^{L/2} \{w\partial x\}x\theta = wL^2\theta/4.$

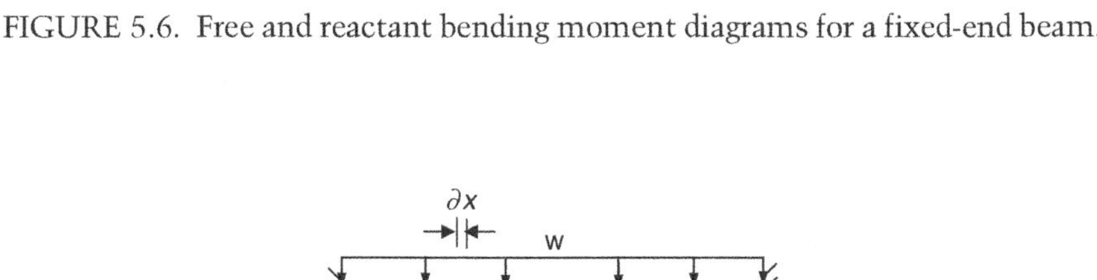

FIGURE 5.6. Free and reactant bending moment diagrams for a fixed-end beam.

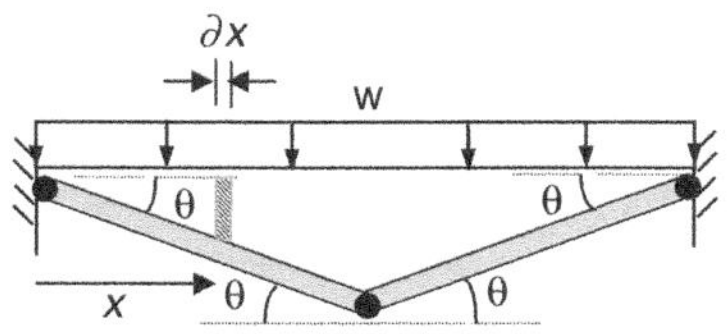

FIGURE 5.7. Collapse mechanism for a fixed-end beam under UDL.

Internal work $= 4M_P\theta$. Because total external work = internal work,

therefore $$w = w_c = \frac{16M_p}{L^2}.$$

5.5.1 Method to Calculate External Work for UDL

Calculation of the external work due to UDL requires integration as demonstrated in the previous example and may become tedious for certain types of structures. Now, consider a UDL of length $L/2$ between two plastic hinges for a straight segment of a beam as shown in Figure 5.8. The equivalent point load for the UDL is $wL/2$ acting at midspan of the segment. The displacement covered by the equivalent point load is $L\theta/4$ and the work done is therefore $(wL/2)(L\theta/4) = \frac{wL^2\theta}{8}$, which is the same as the result by integration. Therefore, for calculation purposes, UDL can be treated as equivalent point loads for any straight segment of the structural members.

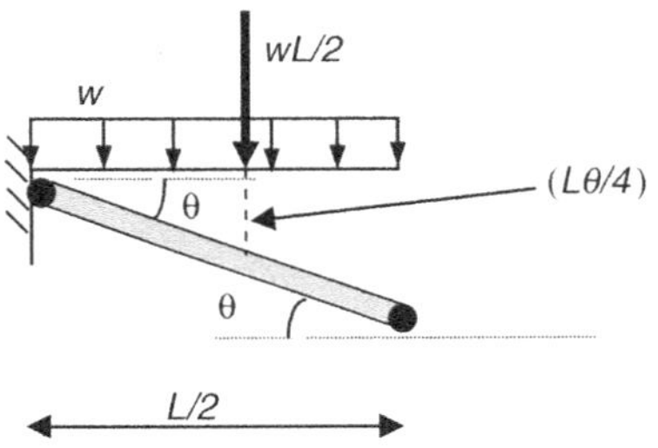

FIGURE 5.8. Treatment of UDL.

Example 5.4 A fixed-end beam ABC is subjected to a UDL of 10α kN/m being applied along AB as shown in Figure 5.9. $M_p = 100$kNm for the beam. Determine the collapse load factor $\alpha = \alpha_c$.

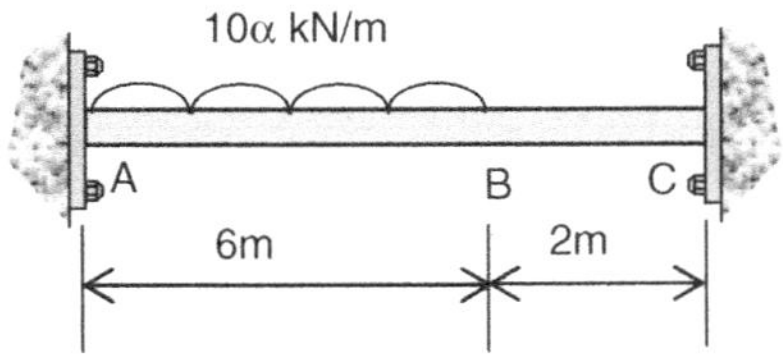

FIGURE 5.9. Example 5.4.

Solution. Three plastic hinges are required to form a collapse mechanism. Assume that the location of the inner plastic hinge is at D at a distance x from A as shown in Figure 5.10. The equivalent point loads for segments AD and DB are used in the work equation. Using the mechanism method, the relationship between θ and β is first established by

$$x\beta = (8 - x)\theta.$$

Internal work $= M_p(2\beta + 2\theta)$.

External work $= 10\alpha x(x/2)\beta + 10\alpha(6 - x)\left(\frac{6 - x}{2} + 2\right)\theta.$

For internal work = external work,

$$M_p = \frac{5\alpha x}{4}(15 - 2x).$$

For maximum bending moment to occur at D,

$$\frac{\partial M_p}{\partial x} = 15 - 4x = 0.$$

Therefore, the plastic hinge occurs at D at $x = 3.75$m.

Hence,

$$M_p = 35.156\alpha \qquad \text{or } \alpha = \alpha_c = 2.84$$

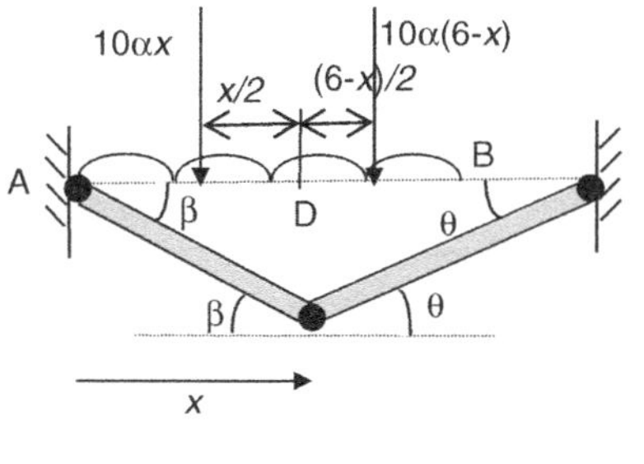

FIGURE 5.10. Equivalent point loads for UDL in Example 5.4.

5.6 Continuous Beams and Frames

In order to examine all possible collapse modes for continuous beams and frames, the concept of partial and complete collapse is introduced in the following section. In particular, partial collapse often occurs in continuous beams and frames upon which multiple loads are applied.

5.6.1 Partial and Complete Collapse

The discussion in the previous sections focused mainly on simple indeterminate structures. Typically, these structures have n degrees of indeterminacy and require $n + 1$ number of plastic hinges to form a collapse mechanism. In such cases, the structures are said to have failed by complete collapse. We define complete collapse as

> *When a structure with* n *degrees of indeterminacy collapses due to the formation of* p *number of plastic hinges where* $p = n + 1$, *the structure fails by complete collapse; in this case, determination of the member forces for the whole structure at collapse is always possible.*

However, partial collapse of a structure can be defined as

> *When a structure with* n *degrees of indeterminacy collapses due to the formation of* p *number of plastic hinges where* $p < n + 1$, *the structure fails by partial collapse; in this case, it may not be possible to determine the member forces for some parts of the structure.*

Structures may fail plastically by complete or partial collapse. In either case, the stiffness of the structure at collapse is zero. For continuous beams and frames where the degree of indeterminacy is large, partial collapse is not uncommon.

5.6.2 Application to Continuous Beams

The procedure for plastic analysis of continuous beams is as follows: identify possible collapse mechanisms, mostly due to partial collapse, in each of the spans; use either the mechanism method or the statical method to find the collapse load or load factor for each collapse mechanism; and identify the critical span, which, when collapse occurs, is the one with smallest collapse load or load factor.

Example 5.5 Determine the collapse load factor $P = P_w$ for the continuous beam shown in Figure 5.11. Plastic moment of the beam is M_p.

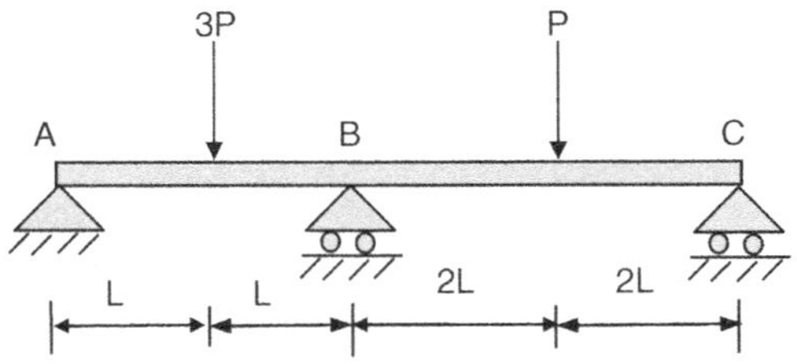

FIGURE 5.11. Plastic collapse of continuous beam.

Solution. The mechanism method is used for the solution.

For left span AB, the plastic hinge occurs at midspan and B as shown in Figure 5.12. The virtual work equation is

$$3P_wL\theta = M_p(3\theta),$$

therefore
$$P_w = \frac{M_p}{L}.$$

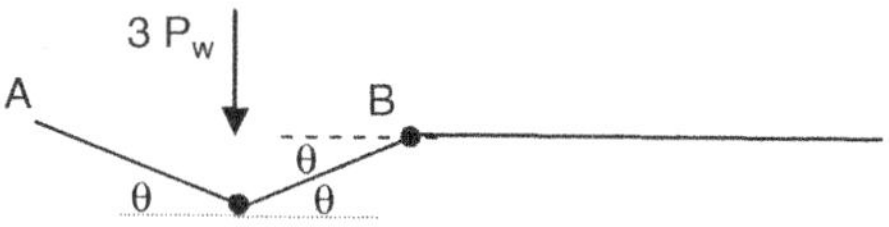

FIGURE 5.12. Collapse mechanism for span AB.

Similarly, for right span BC with two plastic hinges shown in Figure 5.13, the virtual work is

$$P_w2L\theta = M_p(3\theta),$$

therefore
$$P_w = \frac{1.5M_p}{L}.$$

Hence, the left span is critical and $P_w = \dfrac{M_p}{L}$.

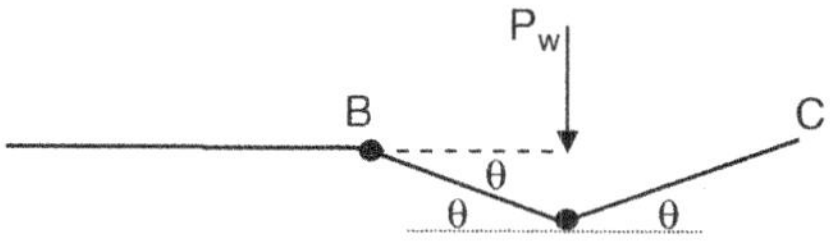

FIGURE 5.13. Collapse mechanism for span BC.

5.6.3 UDL on End Span of a Continuous Beam

In the case of the collapse mechanism at an end span of a continuous beam, it behaves like a propped cantilever due to the formation of two plastic hinges, one within the length and the other at the interior support of the end span. In this case, the exact location of the plastic hinge within the length of the end span needs to be found.

Consider the end span of a continuous beam shown in Figure 5.14. Assume that the internal plastic hinge of the end span is at a distance x from the free end. The collapse mechanism for the end span is also shown in Figure 5.14. The purpose of the analysis is first to minimize

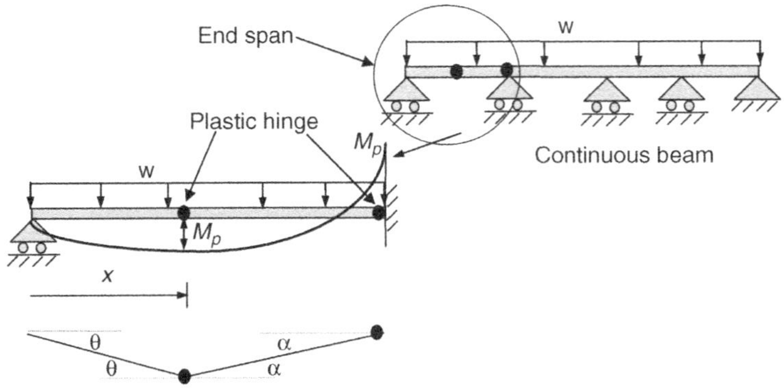

FIGURE 5.14. Collapse mechanism at end span of a continuous beam.

the load w or maximize the bending moment M_p of the internal plastic hinge so that the value of x can be found.

The relationship between the angles of plastic rotation θ and α is

$$\theta x = \alpha(L - x),$$

therefore
$$\alpha = \frac{\theta x}{L - x}.$$

External work $= (wx)\frac{x}{2}\theta + w(L - x)\left(\frac{L - x}{2}\right)\alpha = \frac{wLx}{2}\theta.$

Internal work $= M_p\alpha + M_p(\alpha + \theta) = M_p\left(\frac{L + x}{L - x}\right)\theta.$

External work = Internal work,

therefore
$$w = M_p\left(\frac{2(L + x)}{L(Lx - x^2)}\right) \tag{5.5}$$

For minimum w, $\frac{\delta w}{\delta x} = 0$. It can be proved that if $w = M_p\frac{f_1(x)}{f_2(x)}$, then $\frac{\delta w}{\delta x} = 0$ will lead to the following equation:

$$\frac{f_1(x)}{f_2(x)} = \frac{f'_1(x)}{f'_2(x)} \tag{5.6}$$

where $f'(x)$ represents the first derivative of $f(x)$.

From Equations (5.5) and (5.6),

$$\frac{L + x}{Lx - x^2} = \frac{1}{L - 2x} \qquad \text{giving} \quad x^2 + 2Lx - L^2 = 0,$$

therefore $x = 0.414L.$

Substituting x into Equation (5.5) gives $w = 11.65\,\frac{M_p}{L^2}$.

This is the standard solution of the collapse load for UDL acting on the end span of a continuous beam.

Example 5.6 What is the maximum load factor α that the beam shown in Figure 5.15 can support if $M_p = 93$ kNm?

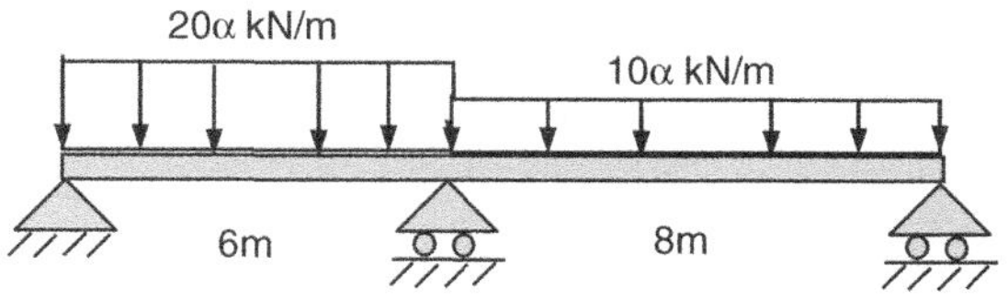

FIGURE 5.15. Example 5.6.

Solution

Left span

$$20\alpha = 11.65\frac{M_p}{L^2} = 11.65\left(\frac{93}{6^2}\right) = 30\text{kN/m},$$

therefore $\alpha = 1.5.$

Right span

$$10\alpha = 11.65\frac{M_p}{L^2} = 11.65\left(\frac{93}{8^2}\right) = 17\text{kN/m},$$

therefore $\alpha = 1.7.$

Hence, the maximum load factor $\alpha = 1.5$

5.6.4 Application to Portal Frames

A portal frame usually involves high degrees of indeterminacy. Therefore, there are always a large number of partial and complete collapse mechanisms (sometimes termed basic mechanisms) that can be combined to form new collapse mechanisms with some plastic hinges becoming elastic (unloading) again. For complex frames, it requires substantial judgment and experience in using this method to identify all possible partial and complete collapse mechanisms.

For simple portal frames, the following types of collapse mechanisms should be identified:

(a) Beam mechanism—when vertical loads are applied to beams and horizontal loads to columns to form partial collapse mechanisms as shown in Figure 5.16.

FIGURE 5.16. Beam mechanisms for beams and columns.

(b) Sway mechanism—when horizontal loads are applied to form complete collapse mechanisms as shown in Figure 5.17.

FIGURE 5.17. Sway mechanism.

(c) Combined mechanism—a combination of beam and sway mechanisms only if unloading occurs to one or more plastic hinges as shown in Figure 5.18.

FIGURE 5.18. Combined mechanism.

Example 5.7 A fixed-base portal frame is subject to a vertical load of 2P and a horizontal load of P shown in Figure 5.19. The length of the rafter is 6L and of the column is 4L. Find the collapse load $P = P_W$.

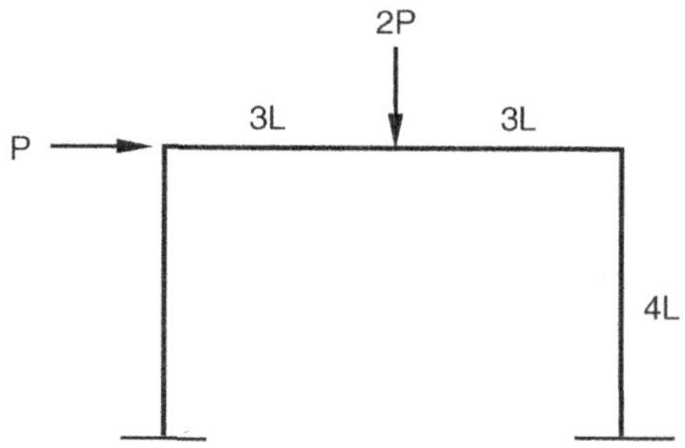

FIGURE 5.19. Example 5.7.

Solution. The portal frame has 3 degrees of indeterminacy. Therefore, a complete collapse mechanism requires four plastic hinges.

(i) Beam mechanism.
From Figure 5.20(i), $2P(3L\theta) = 4M_p\theta$

$$P = \frac{2M_p}{3L}$$

(ii) Sway mechanism.
From Figure 5.20(ii), $P(4L\theta) = 4M_p\theta$

$$P = \frac{M_p}{L}$$

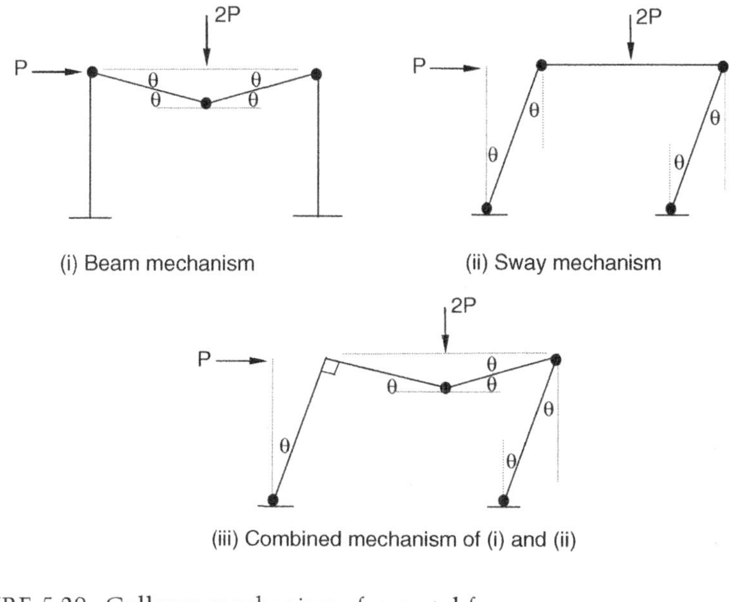

FIGURE 5.20. Collapse mechanisms for portal frame.

(iii) Combined mechanism of (i) and (ii).

From Figure 5.20(iii), $P(4L\theta) + 2P(3L\theta) = 6M_p\theta$

$$P = \frac{3M_p}{5L}$$

Hence, (iii) is critical and $P_w = \dfrac{3M_p}{5L}$.

Example 5.8 A fixed-base portal frame is subject to two horizontal loads of 2P and 3P as shown in Figure 5.21. Find the collapse load $P = P_w$.

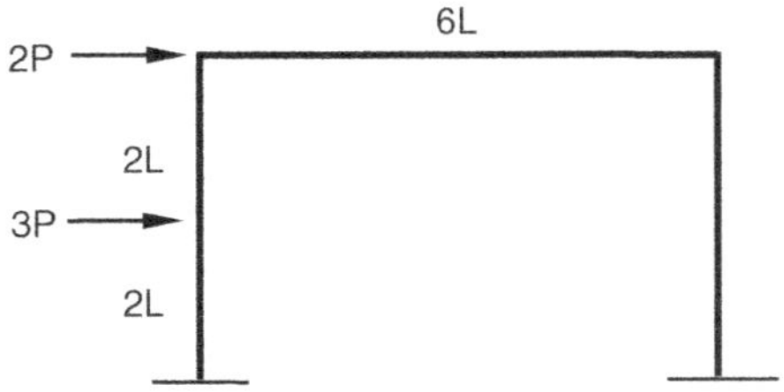

FIGURE 5.21. Example 5.8.

Solution. The three possible collapse mechanisms are shown in Figure 5.22.

(i) Beam mechanism.

$$3P(2L\theta) = 4M_p\theta$$

$$P = \frac{2M_p}{3L}$$

(ii) Sway mechanism.

$$2P(4L\theta) + 3P(2L\theta) = 4M_p\theta$$

$$P = \frac{2M_p}{7L}$$

(iii) Combined mechanism of (i) and (ii).

Relationship between θ and β is $2L\theta = 4L\beta$; hence, $\theta = 2\beta$.

$$2P(2L\theta) + 3P(2L\theta) = M_p(2\theta) + M_p(2\beta)$$

$$P = \frac{3M_p}{10L}$$

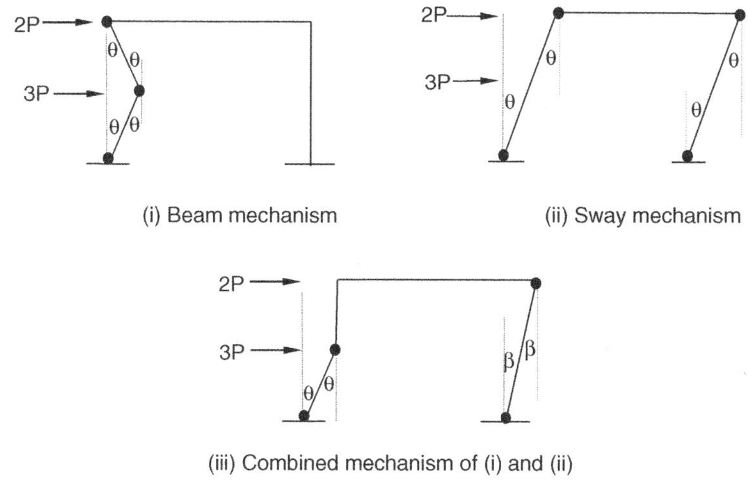

FIGURE 5.22. Collapse mechanisms for Example 5.8.

Hence, mechanism (ii) is critical and $P = P_W = \dfrac{2M_p}{7L}$.

5.7 Calculation of Member Forces at Collapse

While design actions at design load level can be calculated by interpolation in an elastoplastic analysis, it is difficult, if not impossible, to calculate the same using mechanism or statical methods. This is particularly the case for the mechanism method because the behavior of the structure prior to collapse is ignored. The mechanism method is based on a rigid–plastic approach because the structure is assumed to be rigid until collapse occurs. However, when a complete collapse occurs to a structure, it is still possible to calculate the member forces at collapse because the structure is reduced to a determinate one due to the formation of plastic hinges. This statement is not valid for structures failed by partial collapse.

Example 5.9 Calculate the member forces of the structure at collapse for Example 5.7 given that $P_W = \dfrac{3M_p}{5L}$.

Solution. The member forces can be calculated using the appropriate free-body diagrams through the plastic hinges shown in Figure 5.23(i).

For the free-body diagram ED shown in Figure 5.23(ii), take the moment about D:

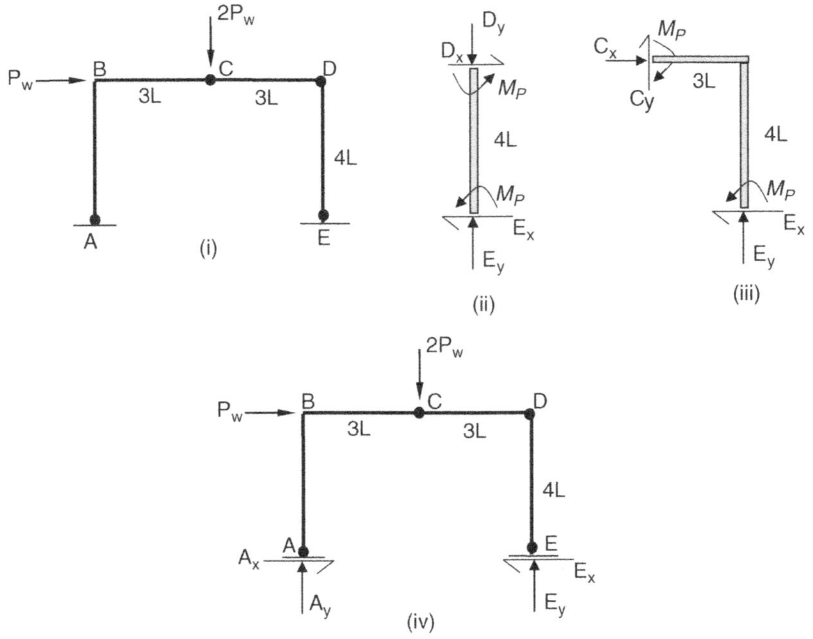

FIGURE 5.23. Free-body diagrams with plastic hinges.

$$M_p + M_p - E_x(4L) = 0$$

$$E_x = \frac{M_p}{2L}$$

Hence,
$$D_x = E_x = \frac{M_p}{2L}.$$

For free-body diagram CDE shown in Figure 5.23(iii), take moment about C:

$$M_p - M_p - E_x(4L) + E_y(3L) = 0$$

$$E_y = \frac{4E_x}{3} = \frac{2M_p}{3L}$$

Also,

$$C_x = E_x = \frac{M_p}{2L}, C_y = -E_y = -\frac{2M_p}{3L}.$$

For free-body diagram ABCDE shown in Figure 5.23(iv), from equilibrium,

$$A_x + P_w - E_x = 0,$$

therefore $$A_x = -\frac{M_p}{10L}.$$

$$A_y - 2P_w + E_y = 0,$$

therefore $$A_y = \frac{8M_p}{15L}.$$

Finally, to verify that the bending moment is elastic at B, a free-body diagram AB shown in Figure 5.24 is chosen and the moment is taken about B:

$$M_p + M + A_x(4L) = 0$$

$$M = \frac{3M_p}{5}$$

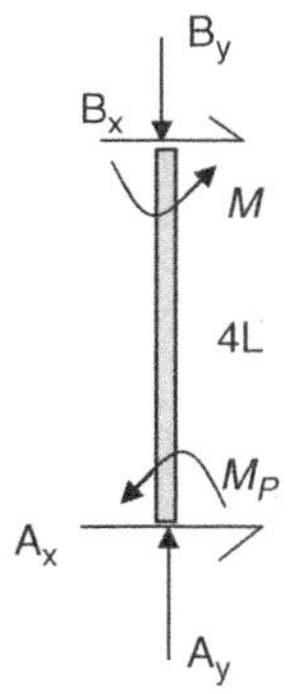

FIGURE 5.24. Free-body diagram for AB.

As expected, the bending moment at B is below M_p and

$$B_x = -A_x = \frac{M_p}{10L}.$$

The bending moment and shear force diagrams for the portal frame are shown in Figure 5.25. The axial forces for the members are $\text{AB} = \frac{8M_p}{15L}$, $BD = \frac{M_p}{2L}$, $\text{ED} = \frac{2M_p}{3L}$.

5.8 Effect of Axial Force on Plastic Collapse Load

As in elastoplastic analysis, the effect of axial force on plastic collapse load can be calculated by successive approximation. The effect of axial force on the plastic moment capacity for each member is calculated at the end of the first iteration, and the reduced moment capacity is used

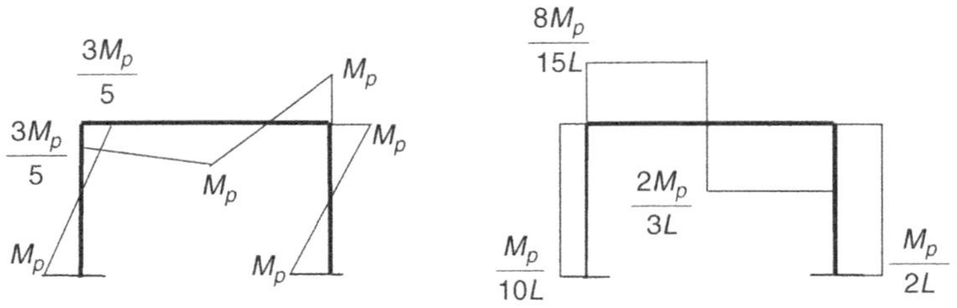

FIGURE 5.25. Bending moment and shear force diagrams.

in the second iteration calculation in which a new collapse load or load factor is obtained. This procedure is repeated until the collapse load converges to a value with sufficient accuracy. The following example demonstrates this process.

Example 5.10 In Example 5.9, if L = 1 m, $M_p = 120$ kNm, $N_p = 320$ kN for all members, determine the collapse load of the structure. The members are made of I section with yield condition given by $m = 1.18(1-\beta)$ where $m = \dfrac{M}{M_P}$ and $\beta = \dfrac{N}{N_P}$ for $\beta > 0.15$, otherwise $m \leq 1$.

Solution. The first iteration gives $P_w = 72$ and the axial forces in the members are AB = 64 kN ($\beta = 0.2$), BD = 60 kN ($\beta = 0.1875$), ED = 80 kN ($\beta = 0.25$). The reduced plastic moments for the members are $(M_{p1})_{AB} = 113.28$ kNm, $(M_{p1})_{BD} = 115.05$ kNm, $(M_{p1})_{ED} = 106.2$ kNm. The subscript "1" indicates the plastic moments at the end of the first iteration. The collapse load factors for the collapse mechanisms are

(i) Beam mechanism:

$$P = \frac{(M_p)_{AB} + 2(M_p)_{BD} + (M_p)_{ED}}{6L}.$$

(ii) Sway mechanism:

$$P = \frac{(M_p)_{AB} + (M_p)_{ED}}{2L}.$$

(iii) Combined mechanism:

$$P = \frac{(M_p)_{AB} + 2(M_p)_{BD} + 3(M_p)_{ED}}{10L}.$$

For the combined mechanism, the axial forces in the members are

$$N_{BD} = \frac{(M_p)_{ED}}{2L}, \quad N_{ED} = \frac{(M_p)_{ED} + (M_p)_{BD}}{3L}, \quad N_{AB} = 2P_w - N_{ED}.$$

TABLE 5.1
Results of solutions by iteration

Iteration	$(M_p)_{AB}$ *(kNm)*	$(M_p)_{BD}$ *(kNm)*	$(M_p)_{ED}$ *(kNm)*	P *(beam mechanism)*	P *(sway mechanism)*	P *(combined mechanism)*	P_w
1	120	120	120	80	120	72	72
2	113.28	115.05	106.2	74.93	109.74	66.20	66.20
3	115.65	118.10	109.0	76.81	112.33	67.89	67.89
4	115.01	117.48	108.10	76.35	111.56	67.43	67.43

Results are given in Table 5.1.

Table 5.1 shows that the collapse load factor P_w converges to a value of 67.43 at the end of the fourth iteration and is about 7% different from that when the bending-axial interaction is not considered.

Problems

5.1. Using the mechanism method, determine the collapse load factor α for the propped cantilever beam ABCD shown in Figure P5.1. $M_p = 80$ kNm.

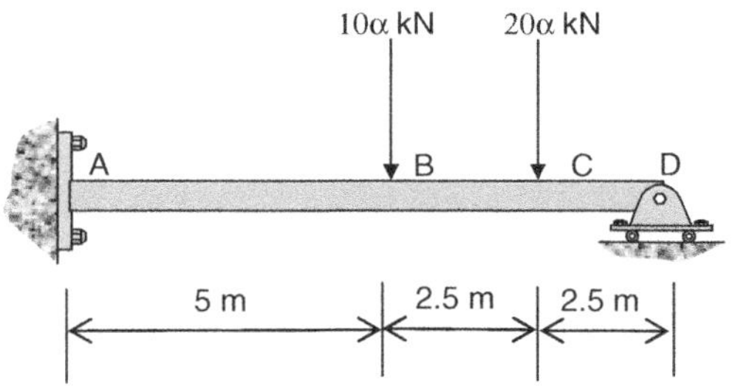

FIGURE P5.1. Problem 5.1.

5.2. Find the minimum plastic moment capacity M_p required for the beam ABC to form a mechanism. Assume that the plastic hinges occur at B and C.

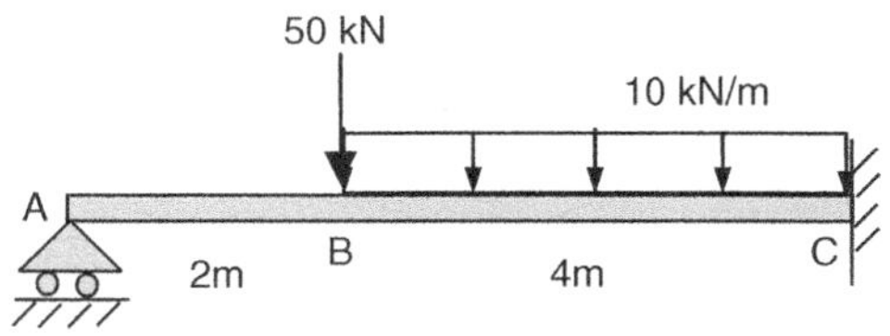

FIGURE P5.2. Problem 5.2.

5.3. Find the common collapse load factor λ applied to the loads for the structure with $M_p = 100$ kNm shown in Figure P5.3.

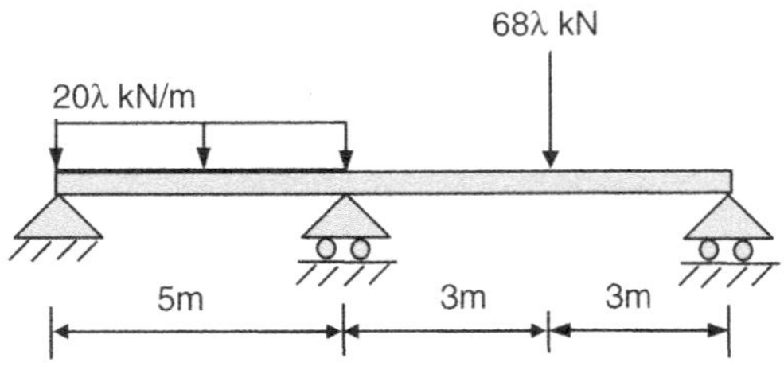

FIGURE P5.3. Problem 5.3.

5.4. Determine the collapse load factor α for the propped cantilever beam ABC subjected to UDL of 10α kN/m along BC shown in Figure P5.4. Locate the plastic hinges at collapse. $M_p = 80$ kNm.

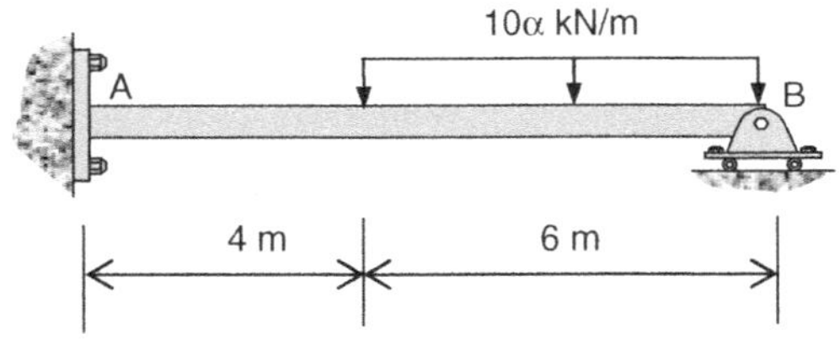

FIGURE P5.4. Propped cantilever under UDL.

5.5. Using the mechanism method, calculate the plastic moment M_p required to support the beam shown in Figure P5.5 before it collapses. Assume that the plastic hinges occur at A, B, and C.

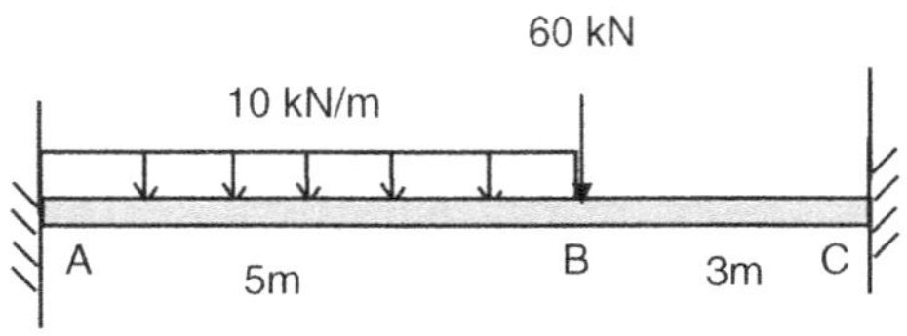

FIGURE P5.5. Fixed-end beam.

5.6. Identify the critical collapse mechanism for the portal frame with one support pinned and the other fixed shown in Figure P5.6 and calculate the common factor P at collapse. Plastic moment $= M_p$.

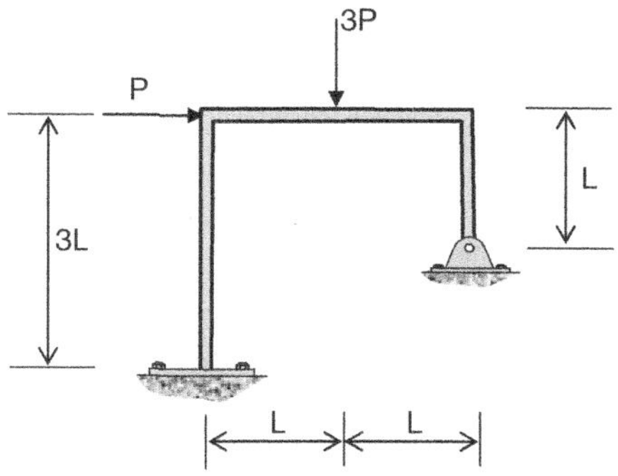

FIGURE P5.6. Problem 5.6.

5.7. Determine the collapse load factor α for the pin-based portal frame shown in Figure P5.7. For all members, $M_p = 200, N_p = 700$, all in consistent units. The members are made of I sections with the yield condition given by $m = 1.18(1-\beta)$ where $m = \dfrac{M}{M_P}$ and $\beta = \dfrac{N}{N_P}$ for $\beta > 0.15$ otherwise $m \leq 1$.

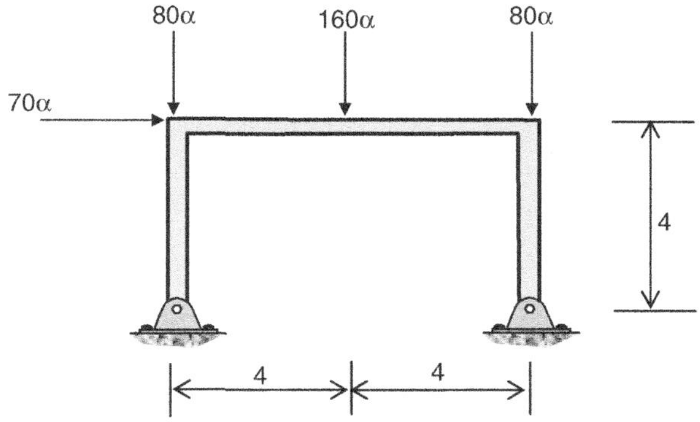

FIGURE P5.7. Pin-based portal frame.

5.8. Determine the value of P at collapse for the column shown in Figure P5.8. The plastic moment of the column is M_p.

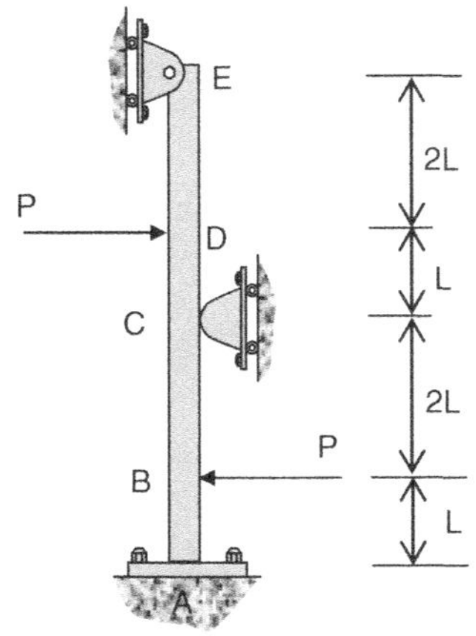

FIGURE P5.8. Problem 5.8.

Bibliography

1. Neal, B. G. (1977). *The plastic methods of structural analysis,* London. Chapman and Hall.
2. Horne, M. R. (1971). *Plastic theory of structures,* Oxford. MIT Press.
3. Beedle, L. S. (1958). *Plastic design of steel frames,* New York. Wiley.
4. Olsen, P. C. (1999). Rigid plastic analysis of plastic frame structures. *Comp. Meth. Appl. Mech. Eng.,* **179**, pp. 19–30.

CHAPTER 6

Limit Analysis by Linear Programming

F. Tin-Loi

6.1 Introduction

We recall that plastic limit analysis is concerned with the problem of finding how "strong" a given structure is. In particular, the aim is to estimate the factor by which the live load component needs to be amplified so that a structural crisis, which takes the form of plastic collapse, occurs. Plastic collapse takes place when the structure is converted into a mechanism by the development of a suitable number and disposition of plastic hinges.

It should be noted that the plastic collapse factor represents one of the most important outcomes of a plastic structural analysis, as it is useful for the reliable and economical safety assessment and design of ductile structures.

As discussed in Chapter 4, this required information can be computed in a step-by-step fashion by following the evolution of the inelastic structural response of a suitably discretized structure to a given loading history. However, such time-stepping analyses are often computationally demanding and may even be unsuitable for preliminary design purposes. Far more appealing would be a direct (nonevolutive) and simple method, namely as carried out through a rigid plastic limit analysis, that is able to provide the plastic collapse load factor quickly.

Classical hand-based calculation methods for limit analysis, as described in Chapter 5 and in some well-known and excellent textbooks,[1–3] are founded on the two fundamental theorems of plasticity. Commonly referred to as the static (safe or lower bound) theorem and the kinematic (unsafe or upper bound) theorem, these notions were developed in the early 1950s and are cumbersome to use, except for the simplest cases of discrete structures for which yielding is governed by a single stress resultant (e.g., by bending moment only instead of by a combination of, say, bending moment and axial force).

Both of these theorems are strongly suggestive of constrained optimization approaches. In fact, Charnes and Greenberg[4] recognized, soon after dual-bound theorems were proposed, that the limit analysis of trusses could be cast as a linear programming (LP) problem. The use of mathematical programming as a conceptual and formal tool for solving a variety of plastic analysis problems, including of course limit analysis, has been the subject of intense research over the last three decades. Some of the key academic events held in the early years on the application of mathematical programming to engineering plasticity include a NATO advanced study course at Waterloo, Canada in 1977; a workshop at Liège, Belgium in 1982; a Euromech Colloquium at Imperial College, London in 1984; a series of lectures at the International Centre for Mechanical Sciences, Udine in 1987; and a conference at the Faculté Polytechnique de Mons, Belgium in 1989. The proceedings of the 1977 Waterloo conference[5] still represent one of the best introductions to this subject.

Unfortunately, despite the fact that such research has clearly highlighted the power and elegance of mathematical programming methods, their use for solving real engineering problems has been sporadic and below expectations. While this may be attributed to a natural tendency for designers to continue using techniques with which they are familiar, the blame also justifiably lies with our university courses that still, in most cases, teach only classical and often tedious techniques involving, for instance, upper bound approximations to the collapse load using an approach based on identification and combination of basic mechanisms.

The primary aim of the present chapter is to fill this gap. We start with brief statements concerning the dual pair of bound theorems of limit analysis that should clearly suggest associated formulations as constrained optimization problems. For simplicity of implementation, we focus on application of the static theorem. Students are then immediately introduced to the solution of some simple examples through application of the LP capability of the popular Microsoft Excel spreadsheet software. We then provide a general description of the discrete plane frame problem with a view to its eventual computer implementation as a MATLAB script. Again the underlying formulation is cast as an LP problem. A note on the optimal plastic design problem concludes the chapter.

6.2 Limit Analysis Theorems as Constrained Optimization Problems

As discussed in Chapter 5, the pair of bound theorems of limit analysis concern statements regarding equilibrium, the mechanism condition, and yield satisfaction (often also called "conformity").

For our purposes, the following are useful statements of the static and kinematic theorems, respectively.

Static limit theorem: An external load computed on the basis of an assumed distribution of stress resultants that satisfy equilibrium and the yield condition is less than or equal to the true collapse load. Hence, among all statically admissible solutions (i.e., sets of such possible stress resultants), the best estimate is the one that maximizes the statically admissible load factor α_s.

It follows that the load factor α_s for any statically admissible solution is a lower bound on the correct load factor α_c for plastic collapse of the structure, i.e., $\alpha_s \leq \alpha_c$.

Kinematic limit theorem: An external load computed on the basis of an assumed mechanism in which the stress resultants satisfy equilibrium is greater than or equal to the true collapse load. Hence, among all kinematically admissible solutions (i.e., possible collapse mechanisms), the best estimate is the one that minimizes the kinematically admissible load factor α_k.

It follows that the load factor α_k for the development of any kinematically admissible solution is an upper bound on the correct load factor α_c for plastic collapse of the structure, i.e., $\alpha_c \leq \alpha_k$.

The following remarks are worth noting:

1. Stated in the just given form, it is immediately obvious that the numerical application of both theorems involved maximization (static theorem) or minimization (kinematic theorem) under the various stated constraints. In particular, the operative form of the static theorem is simply to maximize the load factor (treated as a variable) subject to constraints on equilibrium and yield conformity. The optimum value of this constrained optimization problem, if it exists, will furnish the exact collapse load provided that the discrete model is a proper representation of the actual structure.
2. When the constraints of the optimization problem are linear (as in the case when bending only governs the formation of plastic hinges or when a yield condition involving combined stress resultants has been suitably piecewise linearized), the mathematical programming problem becomes, as is well known, an LP problem. It is then useful to note that if the static LP problem has an optimal solution, then so does the kinematic LP, and moreover the two solutions coincide with the collapse load factor α_c. This uniqueness of the collapse load factor follows from the strong duality property in LP (see, e.g., Chvátal[6]).
3. It cannot be automatically assumed that application of the static (kinematic) theorem on a discrete model of the structure would provide a strict lower (upper) bound. This is only so, as indicated

earlier, if the model is a true representation of the actual structure. For instance, any model that involves yield conformity checks at only some sections, leaving out some critical ones that do yield, is unlikely to provide a lower bound. Similarly, if a nonlinear yield surface has been approximated as an inscribed piecewise linear one, the strength of the plastic zone will be underestimated, and it is highly likely that the collapse load will also be underestimated.

6.3 Spreadsheet Solution of Simple Limit Analysis Problems

Our purpose in this section is to formulate and solve, as LP problems, some simple beam and frame problems for which bending governs only the formation of plastic hinges. For this purpose, we use the static theorem and recall that its application involves the following steps.

(a) Identify the discrete locations in the structure that are deemed to be critical, that is, locations at which plastic flexural hinges may occur.

(b) Develop appropriate equilibrium equations for the bending moment at each of the chosen critical sections. This can be achieved easily in the same fashion as the well-known step in the flexibility analysis of structures. In essence, the actual equilibrium distribution of bending moments can be expressed as the sum of the bending moments on a primary (suitably released) structure subjected to the applied loads and of an unknown self-stress moment distribution. This can be written formally as

$$\mathbf{m} = \alpha \mathbf{B}_0 \mathbf{f} + \mathbf{B}_1 \mathbf{x} \tag{6.1}$$

where vector $\mathbf{m}$ collects the bending moments at all critical sections, $\mathbf{f}$ are the given known loads (governed by the load multiplier α), $\mathbf{x}$ is a vector of redundants (releases), and quantities $\mathbf{B}_0$ and $\mathbf{B}_1$ are referred to, respectively, as the basic and redundant load matrices. For simple structures, such as the ones considered in this chapter, the redundants can be chosen by inspection and the necessary matrices can then be easily calculated manually by statics. Accurate and efficient techniques exist for automatic selection of $\mathbf{x}$, and consequent generation of $\mathbf{B}_0$ and $\mathbf{B}_1$, but these are not to be discussed in this book.

(c) List the yield conditions that apply at each of the critical sections. If we assume, for simplicity, that the plastic moment capacities $\mathbf{M}_p$ in both positive and negative bending are equal at all these sections, then these conditions are

$$-\mathbf{M}_p \leq \mathbf{m} \leq \mathbf{M}_p. \tag{6.2}$$

(d) Use Equations (6.1) and (6.2) as constraints and set up the static limit analysis problem as the following LP problem:

$$\left.\begin{array}{ll} \text{maximize} & \alpha \\ \text{subject to} & \mathbf{m} = \alpha \mathbf{B}_0 \mathbf{f} + \mathbf{B}_1 \mathbf{x}, \\ & -\mathbf{M}_p \leq \mathbf{m} \leq \mathbf{M}_p. \end{array}\right\} \tag{6.3}$$

(e) Input LP data for solution by a selected LP solver. This section outlines use of the LP solver facility built into the Excel spreadsheet.

Example 6.1: Simple Propped Cantilever

This first example is that of a simple propped cantilever (Figure 6.1a) of constant plastic moment capacity $M_p = 10$.

This is a problem that can be solved easily by hand calculations. However, our purpose is to clarify how various data can be obtained for setting up the appropriate LP problem and also to detail how Excel can be used to obtain the optimal solution.

The structure is onefold statically indeterminate and we choose the vertical reaction r (assumed to be positive up) at the propped support as the redundant (Figure 6.1b). The bending moment distributions corresponding to the primary (released) structure and to the redundant are as shown, respectively, in Figures 6.1c and 6.1d; the assigned signs are consistent but arbitrary (with sagging moments assumed to be positive). These distributions can be used to set up the equilibrium conditions $\mathbf{m} = \alpha \mathbf{B}_0 \mathbf{f} + \mathbf{B}_1 \mathbf{x}$ once the locations of critical sections have been decided upon. It should also be noted that

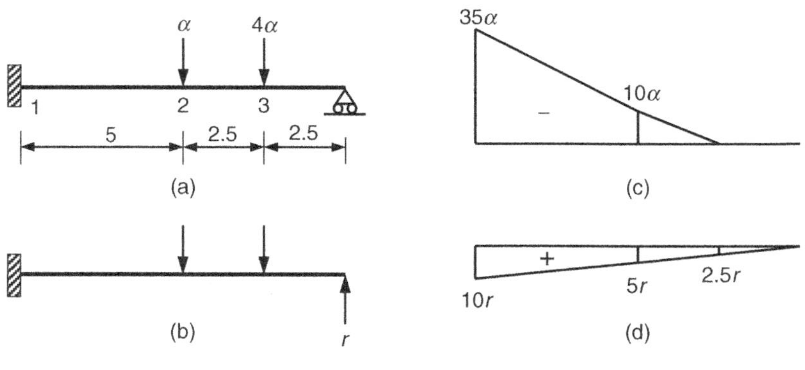

FIGURE 6.1. Example 6.1: Simple propped cantilever.

the entries for $\mathbf{B}_0\mathbf{f}$ are directly available from the bending moment distribution shown in Figure 6.1c.

Let us assume that there are three potential hinge locations, indicated by the numbers 1, 2, and 3 in Figure 6.1a.

It is now a simple matter to set up the equilibrium conditions for these locations as follows:

$$\begin{bmatrix} m_1 \\ m_2 \\ m_3 \end{bmatrix} = \begin{bmatrix} -35 \\ -10 \\ 0 \end{bmatrix} \alpha + \begin{bmatrix} 10 \\ 5 \\ 2.5 \end{bmatrix} r$$

where the subscripts used in conjunction with the m values designate the critical section number. It is important to remember that variable r for LP purposes is a "free" variable (namely, $-\infty \leq r \leq \infty$). Also, α, being the only term in the objective function that is to be maximized, can be similarly treated as a free variable.

The yield conditions for each of the three sections can be written explicitly as

$$\begin{bmatrix} -10 \\ -10 \\ -10 \end{bmatrix} \leq \begin{bmatrix} m_1 \\ m_2 \\ m_3 \end{bmatrix} \leq \begin{bmatrix} 10 \\ 10 \\ 10 \end{bmatrix}.$$

Before we start entering required data into Excel for solution by its linear optimizer, it is useful to reduce the number of variables by eliminating m_1, m_2, and m_3. This can be achieved by combining the m expressions with their corresponding yield constraints. After this has been carried out, a complete specification for our limit analysis problem [Equation (6.3)] in LP format is then as follows:

Objective: max α
Variables: α, r
Constraint1: $-35\alpha + 10r \leq 10$
Constraint2: $-35\alpha + 10r \geq -10$
Constraint3: $-10\alpha + 5r \leq 10$
Constraint4: $-10\alpha + 5r \geq -10$
Constraint5: $2.5r \leq 10$
Constraint6: $2.5r \geq -10$

As expected, there are six constraints (or two per critical section representing positive and negative yielding). Also, note that in Excel, we do not need, as shown earlier, to explicitly provide bounds for the free variables α and r.

The following now describes in some detail how Excel is used to solve the LP problem.

(a) *Check if the solver add-in is present.* Open Excel and check if the **Solver** command is present on the **Tools** menu. If not, it should be added in by using the **Add-ins** menu, selecting **Solver Add-in,** and clicking **OK**. If the **Solver Add-in** is not currently installed, you will be asked whether you want to install it and you can proceed to add it in.

(b) *Prepare an Excel sheet for data entry and processing.* While it is not essential, it would be useful to label your worksheet for ease of checking for data entry errors and to have a clearer picture as to which variable and constraint values are being referenced. The layout largely depends on the user's preference. We will adopt a simple and largely uncluttered worksheet. We proceed as follows:

- In cell A1, provide a short description of the example.
- Provide names of the two variables in cells A3 and A4.
- Provide names of the six constraints in cells A6 to A11.
- Enter RHS values for all constraints in cells B6 to B11 and label the column at cell B5. This step and all of the aforementioned steps are mandatory, as RHS values could be entered directly after naming the **Solver**.
- All variable and constraint function values will be kept in the appropriate C column, namely C3 to represent α, C4 to represent r, C6 to contain the explicit constraint function for Constraint1, etc.
- We now enter these constraint functions: click on cell C6 and define it by entering the formula = –35*C3 + 10*C13, click on cell C7 and define it by = –35*C3 + 10*C13, etc. until the last cell, C11, has been defined.

At this stage, the spreadsheet should appear as in Figure 6.2.

(c) *Enter **Solver** parameters.* Complete the following steps to enter the LP information that the solver needs to compute the collapse load multiplier:

- On the **Tools** menu, click **Solver**. You should see the box in Figure 6.3.
- In the **Set Target Cell** box, enter the cell reference for the objective function. In our case, it coincides with variable α or cell C3.
- Make sure that **Max** is selected since we are maximizing.
- In the **By Changing Cells** box, enter the cell references for each of the two variables. In our case, enter C3:C4 (by clicking and dragging on the appropriate cells in typical Excel fashion).
- In the **Subject to the Constraints** box, click the **Add** option. After an **Add Constraint** dialog appears, enter the **Cell Reference** for the first constraint (i.e., C6), the appropriate type of constraint

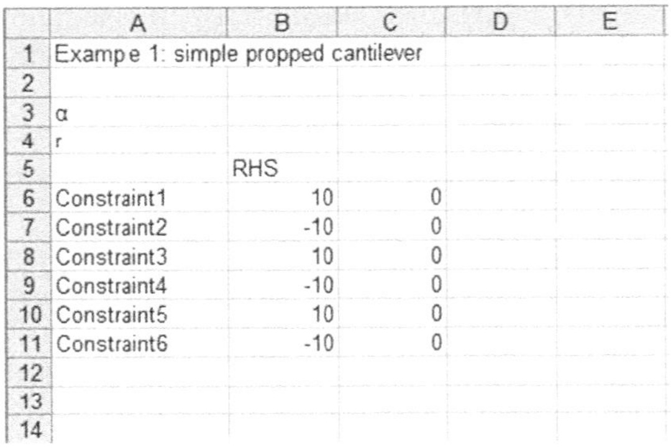

FIGURE 6.2. Example 6.1: Spreadsheet layout for LP problem.

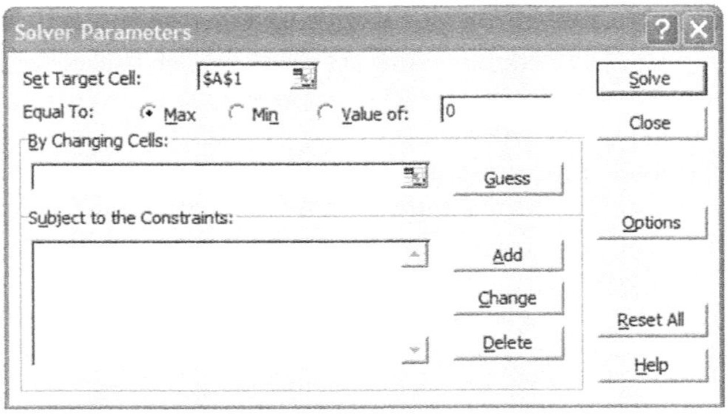

FIGURE 6.3. Example 6.1: Solver parameters box.

(i.e., ≤), and the RHS of the constraint (i.e., B6). The result should be as in Figure 6.4.

- Click **Add** (or **OK**) and repeat for each of the remaining five constraints. The final box should be as in Figure 6.5.

(d) *Solve the LP problem and obtain the output.*

- We are now ready to solve the LP problem. Simply click **Solve**. You will see the box in Figure 6.6.

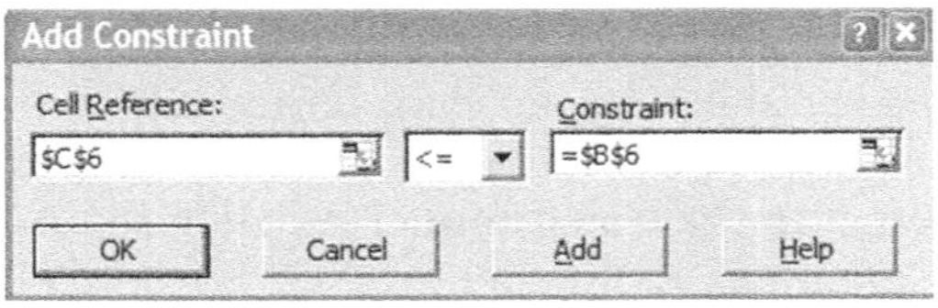

FIGURE 6.4. Example 6.1: Add constraint box.

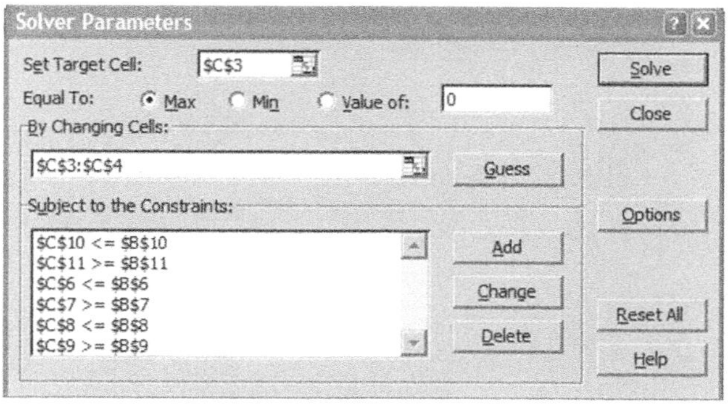

FIGURE 6.5. Example 6.1: Completed solver parameters box.

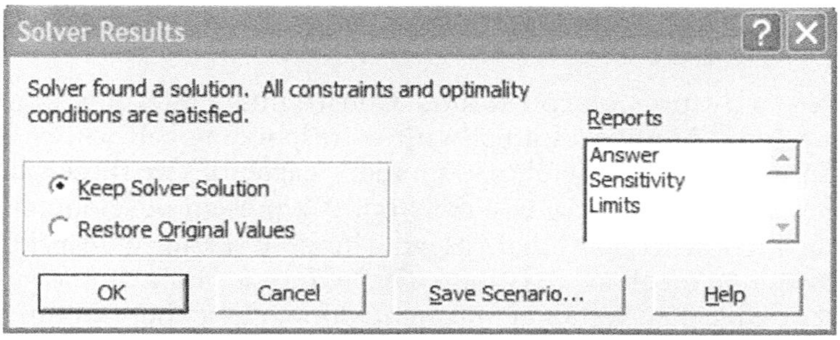

FIGURE 6.6. Example 6.1: Result of solve.

- If you wish to keep the solution values on the worksheet, click **Keep Solver Solution**. This means that the values for your variables have been changed with the corresponding changes to the cells defining objective function and constraint functions. Because we may wish to test this model further, we revert back to the original values by simply clicking on **Restore Original Values**.
- Finally, you can choose the type of report you wish to generate in the **Reports** box. In our case, we wish to have a summary of the results. Hence click **Answer** and then **OK**. You will notice that a new sheet marked **Answer Report 1** has been generated.
- The report is given in Figure 6.7.

(e) *Interpreting the results.* The report allows easy interpretation of our limit analysis run.

In summary, it indicates that the limit load multiplier $\alpha_c = 1.4286$, the redundant $r = 4$, and the mechanism produced

Target Cell (Max)

Cell	Name	Original Value	Final Value
\$C\$3	α	0	1.428571429

Adjustable Cells

Cell	Name	Original Value	Final Value
\$C\$3	α	0	1.428571429
\$C\$4	r	0	4

Constraints

Cell	Name	Cell Value	Formula	Status	Slack
\$C\$6	Constraint1	-10	\$C\$6<=\$B\$6	Not Binding	20
\$C\$7	Constraint2	-10	\$C\$7>=\$B\$7	Binding	0
\$C\$8	Constraint3	5.714285714	\$C\$8<=\$B\$8	Not Binding	4.285714286
\$C\$9	Constraint4	5.714285714	\$C\$9>=\$B\$9	Not Binding	15.71428571
\$C\$10	Constraint5	10	\$C\$10<=\$B\$10	Binding	0
\$C\$11	Constraint6	10	\$C\$11>=\$B\$11	Not Binding	20

FIGURE 6.7. Example 6.1: Results report.

involves activation of constraints Constraint2 and Constraint5 in view of their "binding" status (with corresponding cell values at the limit indicating that the plastic moment capacities at these sections have been reached). These two constraints represent development of a hogging hinge at section 1 and a sagging hinge at section 3, respectively. This particular mechanism is shown in Figure 6.8. Unfortunately, Excel cannot provide dual values at optimality; in our case these would correspond to hinge rotations as the constraints for the static problem relate to yield conditions. The interpretation of optimal dual values is clarified in later sections of this chapter.

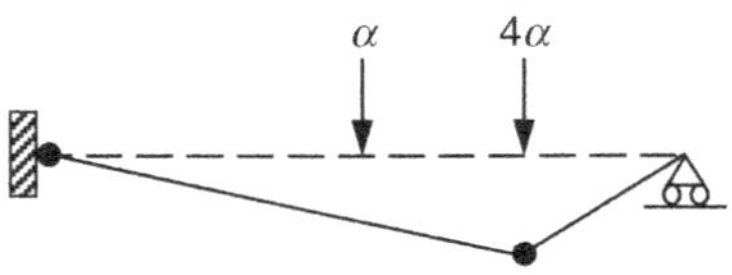

FIGURE 6.8. Example 6.1: Mechanism.

Example 6.2: Pinned Gable Frame

We now provide a second example to reinforce how the static limit analysis problem can be set up for Excel solution. Fewer details, as compared to Example 6.1, are provided; these are left to the student to work out.

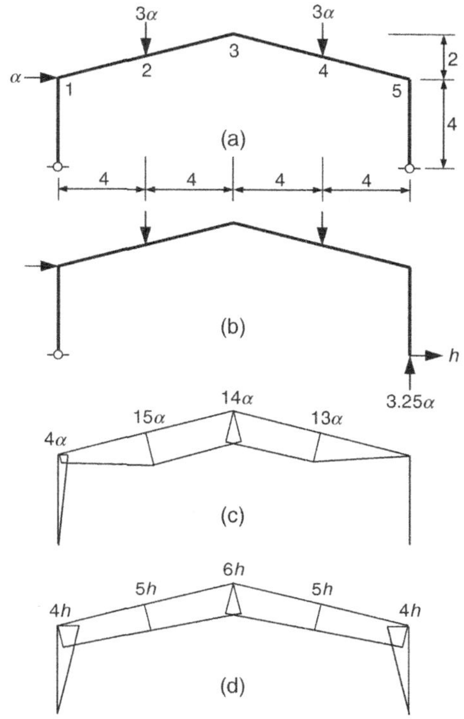

FIGURE 6.9. Example 6.2: Pinned gable frame.

The structure, which is a simple gable frame with pinned supports (Figure 6.9a), has been deliberately selected because it traditionally poses various problems for students who try to solve it manually by application of the mechanism approach. Many find it difficult to identify critical mechanisms and to calculate appropriate hinge rotations for use in the traditional virtual work approach.

For our static LP approach, we select the horizontal reaction at the right-hand support as the redundant (Figure 6.9b); the vertical reaction is 3.25α, as can be verified by taking moment equilibrium about the left-hand support. Simple statics will furnish the bending moments corresponding to applied loads (Figure 6.9c) and redundant (Figure 6.9d); bending moments are assumed to be positive if they cause tension on the inner faces of members.

Assuming, as before, a structure of constant plastic moment capacity $M_p = 10$, we can set up an LP problem in the same format as in Example 1, namely,

Objective: $\max \alpha$
Variables: α, h
Constraint1: $4\alpha + 4h \leq 10$
Constraint2: $4\alpha + 4h \geq -10$
Constraint3: $15\alpha + 5h \leq 10$
Constraint4: $15\alpha + 5h \geq -10$
Constraint5: $14\alpha + 6h \leq 10$
Constraint6: $14\alpha + 6h \geq -10$
Constraint7: $13\alpha + 5h \leq 10$
Constraint8: $13\alpha + 5h \geq -10$
Constraint9: $4h \leq 10$
Constraint10: $4h \geq -10$

After entering and solving this LP problem in Excel, we obtain the solution report shown in Figure 6.10.

As the results report shows, the optimal solution is $\alpha_c = 1.5$, the redundant $h = -2.5$ (the negative sign indicates that this horizontal reaction acts in the opposite direction to the one initially assumed), and the mechanism involves formation of a positive moment hinge at section 2 and of a negative moment hinge at section 5. The fact that the two plastic hinges are formed is not surprising, as the original structure is onefold indeterminate. This information allows us to draw the collapse mechanism, which is displayed in Figure 6.11.

Target Cell (Max)

Cell	Name	Original Value	Final Value
\$C\$3	α	0	1.5

Adjustable Cells

Cell	Name	Original Value	Final Value
\$C\$3	α	0	1.5
\$C\$4	h	0	-2.5

Constraints

Cell	Name	Cell Value	Formula	Status	Slack
\$C\$6	Constraint1	-4	\$C\$6<=\$B\$6	Not Binding	14
\$C\$7	Constraint2	-4	\$C\$7>=\$B\$7	Not Binding	6
\$C\$8	Constraint3	10	\$C\$8<=\$B\$8	Binding	0
\$C\$9	Constraint4	10	\$C\$9>=\$B\$9	Not Binding	20
\$C\$10	Constraint5	6	\$C\$10<=\$B\$10	Not Binding	4
\$C\$11	Constraint6	6	\$C\$11>=\$B\$11	Not Binding	16
\$C\$12	Constraint7	7	\$C\$12<=\$B\$12	Not Binding	3
\$C\$13	Constraint8	7	\$C\$13>=\$B\$13	Not Binding	17
\$C\$14	Constraint9	-10	\$C\$14<=\$B\$14	Not Binding	20
\$C\$15	Constraint10	-10	\$C\$15>=\$B\$15	Binding	0

FIGURE 6.10. Example 6.2: Results report.

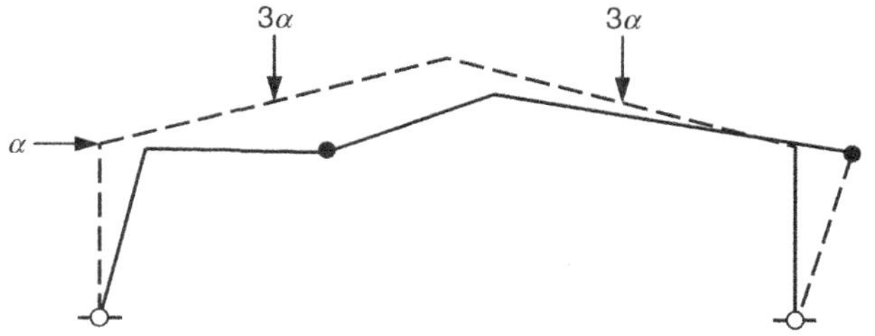

FIGURE 6.11. Example 6.2: Mechanism.

6.4 General Description of the Discrete Plane Frame Problem

It should be clear by now that the spreadsheet LP approach described in the previous section eliminates the often tedious and sometimes difficult process involved in the hand calculation of trial mechanisms. A disadvantage, however, lies in the fact that, except for simple structures for which yielding is governed by a single stress resultant, the hand generation of equilibrium and yield conditions can become a formidable task.

The aim of this section is to describe how the limit analysis for large plane frame problems, involving combined bending-axial force plastic hinges (as well as pure moment hinges), can be systematically set up for automatic generation via a spreadsheet or other popular modeling environment, such as MATLAB.

The structural idealization described is based on the familiar static-nodal system representation invariably used in finite element modeling. As described previously, we adopt the static LP approach to find the limit load multiplier.

6.4.1 Mathematical Model and Governing Relations

Structural Model

The following basic assumptions are made:

(a) The structure is plane and is made up of straight prismatic finite elements connected at nodes.
(b) Plastic deformations are assumed to be lumped at the ends of each element while the element itself is considered to be rigid.
(c) A generalized plastic hinge model, extended to include axial effects, is assumed valid.
(d) Classical perfect plasticity conditions apply so that plastic deformations are consistent with the normality rule.
(e) Displacements are small so that equilibrium equations refer to the initial undeformed structure geometry.

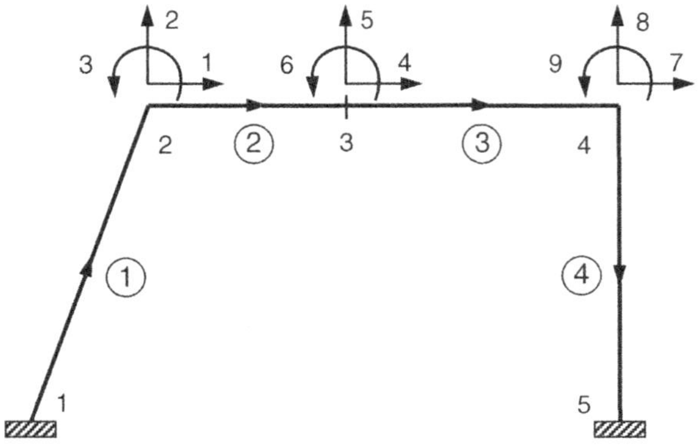

FIGURE 6.12. Structural model of a plane frame.

The frame is idealized as a number of elements that are interconnected at a discrete number of nodes. The discretization is made on the basis of known or assumed locations at which hinges may form. As an example, the planar frame shown in Figure 6.12 has been modeled using four elements and five nodes. All external forces are reduced to act on these nodes in the form of concentrated forces. Thus, for the model shown, the equivalent nodal load vector $\alpha\mathbf{f}$ has a dimension (9×1), consistent with the 9 degree of freedom model. As usual, α is the live load multiplier and $\mathbf{f}$ is known. To simplify the notation, let $\mathbf{p} = \alpha\mathbf{f}$. Member axes are also shown with, for instance, element 1 oriented in the node 1 to node 2 direction.

Each finite element is described in terms of three independent (or so-called "natural") stress resultants (thus eliminating rigid body motions). The particular choice adopted is as shown in Figure 6.13. It consists of three normalized stress resultants, namely axial force and bending moment at the "start" node j of the generic ith element and a bending moment at the "end" node k of the element. These nondimensional generalized forces, collected in vector $\mathbf{s}^i$, are represented, respectively, by s_1^i, s_2^i, and s_3^i. More explicitly, $s_1^i = n_j/N_p^i$,

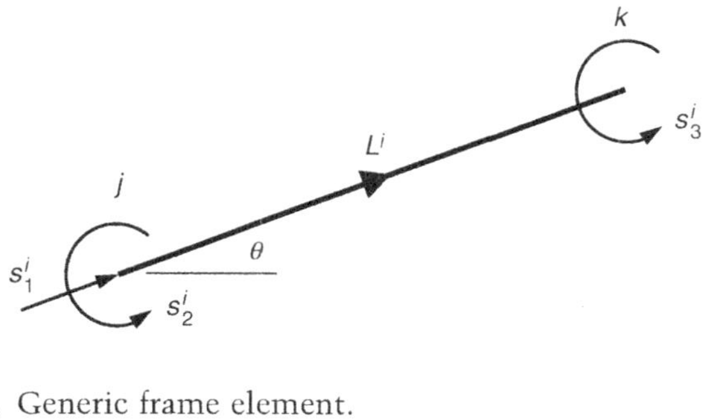

FIGURE 6.13. Generic frame element.

$s_2^i = m_j/M_p^i$ and $s_3^i = m_k/M_p^i$, where n is the axial force, m is the bending moment, N_p is the pure axial plastic capacity of the member, and M_p is its pure moment plastic capacity; various subscripts and superscripts used are self-explanatory.

Equilibrium

We are now in a position to establish the equilibrium relations for the structure. For this purpose, consider the generic frame element i of length L^i oriented at an angle θ to the horizontal, as in Figure 6.14. The three selected independent stress resultants $\mathbf{s}^i$ are shown together with the remaining three stress resultants required to self-equilibrate the member. Also indicated are the corresponding proportions of the applied loads, collected in the (6×1) vector $\mathbf{p}^i$, on end nodes j and k required to maintain nodal equilibrium. More explicitly,

$$\mathbf{p}^{iT} = [p_{j1}, p_{j2}, p_{j3}, p_{k1}, p_{k2}, p_{k3}].$$

Equilibrium at the element level can be simply established by writing the six nodal equilibrium equations. For node j, the conditions for horizontal, vertical, and moment equilibrium can be written, respectively, as

$$\Sigma \mathrm{H} = 0 = p_{j1} - s_1^i \cos\theta + \frac{s_2^i + s_3^i}{L^i} \sin\theta,$$

$$\Sigma \mathrm{V} = 0 = p_{j2} - s_1^i \sin\theta - \frac{s_2^i + s_3^i}{L^i} \cos\theta,$$

$$\Sigma \mathrm{M} = 0 = p_{j3} - s_2^i,$$

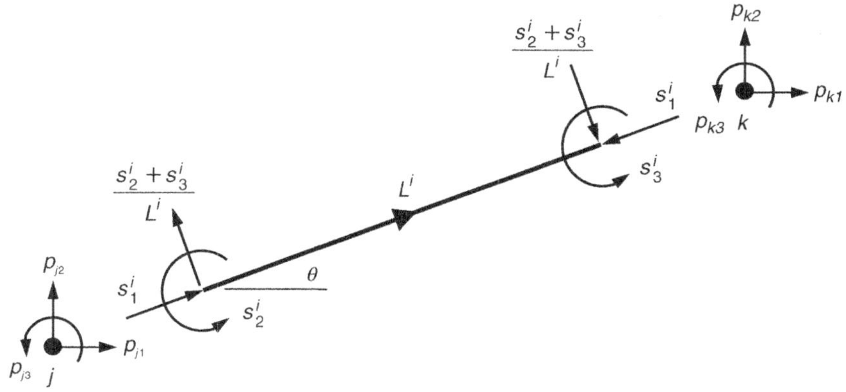

FIGURE 6.14. Self-equilibrated element and corresponding nodal loads.

and similarly for node k as

$$\Sigma H = 0 = p_{k1} + s_1^i \cos\theta - \frac{s_2^i + s_3^i}{L^i} \sin\theta,$$

$$\Sigma V = 0 = p_{k2} + s_1^i \sin\theta + \frac{s_2^i + s_3^i}{L^i} \cos\theta,$$

$$\Sigma M = 0 = p_{k3} - s_3^i.$$

These relations can be compactly expressed as

$$\mathbf{B}^i \mathbf{s}^i = \mathbf{p}^i \tag{6.4}$$

where $\mathbf{B}^i$ represents the (6×3) element equilibrium matrix

$$\mathbf{B}^i = \begin{bmatrix} \cos\theta & -\sin\theta/L^i & -\sin\theta/L^i \\ \sin\theta & \cos\theta/L^i & \cos\theta/L^i \\ 0 & 1 & 0 \\ -\cos\theta & \sin\theta/L^i & \sin\theta/L^i \\ -\sin\theta & -\cos\theta/L^i & -\cos\theta/L^i \\ 0 & 0 & 1 \end{bmatrix}.$$

We are now in a position to generate the equilibrium conditions at the structure level. This structure equilibrium can be written as

$$\mathbf{B}s = \mathbf{p} \tag{6.5}$$

where, for a structure discretized into n elements and d degrees of freedom, $\mathbf{B}$ is the $(d \times 3n)$ structure equilibrium matrix, $\mathbf{s}$ is the $(3n \times 1)$ vector of all stress resultants collected in the traditional concatenated form $\mathbf{s}^T = [\mathbf{s}^{1T} \ldots \mathbf{s}^{nT}]$, and $\mathbf{p}^T = [p_1 \ldots p_d]$.

$\mathbf{B}$ can be assembled from element $\mathbf{B}^i$ matrices through a process very similar to the assembly of stiffness matrices. In essence, it uses location or freedom vectors that contain information relating to where elements of $\mathbf{B}^i$ need to be inserted into $\mathbf{B}$. For instance, element 4 of the frame shown in Figure 6.12 has the location vector [7 8 9 0 0 0] so that the entire first row of $\mathbf{B}^4$ is placed in the seventh row of the (9×12) matrix $\mathbf{B}$ in the columns appropriate to frame element 4, namely columns 10, 11, and 12; the second row of $\mathbf{B}^4$ is placed in the eighth row of $\mathbf{B}$ in the same columns; the third row of $\mathbf{B}^4$ is placed in the ninth row of $\mathbf{B}$ in the same columns; and the three remaining rows of $\mathbf{B}$ are set to zeros since the corresponding location vector entries are zeros.

Piecewise Linear Yield Conditions

As indicated previously, a generalized plastic hinge can form at any of the member ends. For the generic member i (Figure 6.13), a hinge can thus form at node j through the combined axial-bending interaction of the active stress resultants (s_1^i, s_2^i) and another hinge at node k through the effect of (s_1^i, s_3^i). Of course, the description presented here also includes, as a special case, the more familiar and simpler case of hinge formation through bending action only.

The yield conformity condition is typically nonlinear. However, for our purposes, we will adopt a suitably piecewise linearized representation of such yield surfaces. This is computationally advantageous, as the underlying discrete limit analysis problem can be cast as an LP (rather than nonlinear programming) problem.

As an example, consider the case of a solid rectangular section under combined axial force and bending. For node j of element i, a simple eight-hyperplane lower bound approximation to the well-known nonlinear condition (e.g., Massonnet and Save[2])

$$(s_1^i)^2 + |s_2^i| \leq 1$$

is as shown in Figure 6.15.

The eight yield conditions (one for each hyperplane) that govern the linearized yield locus can be written explicitly, in order of the hyperplane numbering shown in Figure 6.15, as follows:

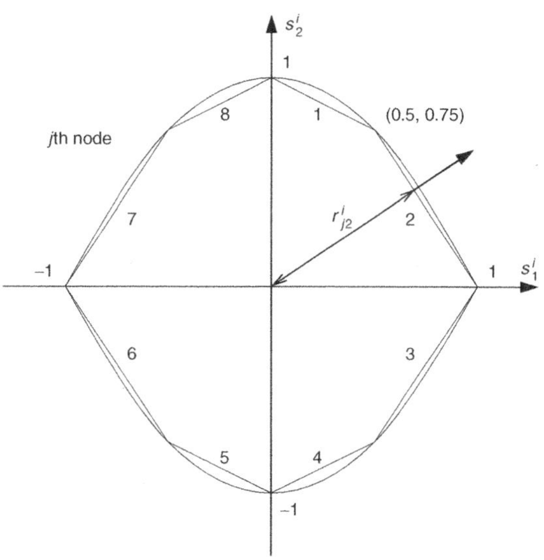

FIGURE 6.15. Piecewise linear approximation of nonlinear yield locus.

$$\begin{aligned}
\ell_1 s_1^i + m_1 s_2^i &\leq r_{j1}^i,\\
\ell_2 s_1^i + m_2 s_2^i &\leq r_{j2}^i,\\
\ell_2 s_1^i - m_2 s_2^i &\leq r_{j2}^i,\\
\ell_1 s_1^i - m_1 s_2^i &\leq r_{j1}^i,\\
-\ell_1 s_1^i - m_1 s_2^i &\leq r_{j1}^i,\\
-\ell_2 s_1^i - m_2 s_2^i &\leq r_{j2}^i,\\
-\ell_2 s_1^i + m_2 s_2^i &\leq r_{j2}^i,\\
-\ell_1 s_1^i + m_1 s_2^i &\leq r_{j1}^i
\end{aligned}$$

where ($\ell_1 = 1/\sqrt{5}$, $m_1 = 2/\sqrt{5}$) are the direction cosines of the normal to hyperplane 1, ($\ell_2 = 3/\sqrt{13}, m_2 = 2/\sqrt{13}$) are the direction cosines of the normal to hyperplane 2, r_{j1}^i is the perpendicular distance of hyperplane 1 from the origin, and r_{j2}^i is the perpendicular distance of hyperplane 2 from the origin.

Before providing a compact vector–matrix representation of these relations, it would be worthwhile to give a geometric explanation of the yield condition pertaining to one hyperplane. Consider the stress resultant OS with components (a, b) and a single hyperplane, as shown in Figure 6.16.

Geometrically, the yield condition is expressed as OB $\leq$ OH. Thus, when OB, the projection of OS on OH, is less than OH, the stress point lies in the rigid region. However, when OB = OH, the stress point is located on the yield hyperplane. Now, OB = OA + AB, and because OA = $a\cos\beta$, AB = $b\sin\beta$, and OH = r, then the condition OB $\leq$ OH can be written as $a\cos\beta + b\sin\beta \leq r$, which is precisely

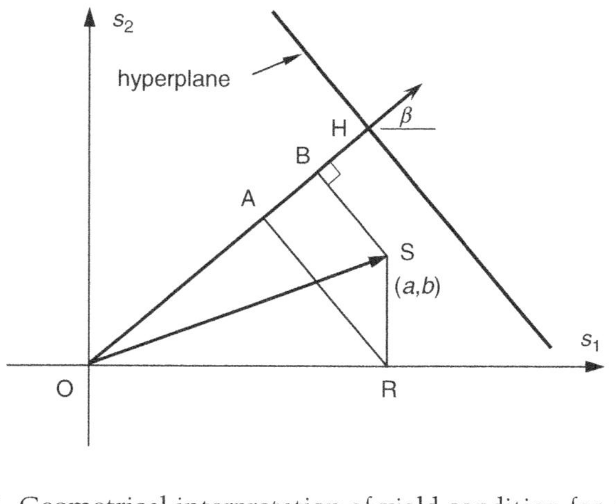

FIGURE 6.16. Geometrical interpretation of yield condition for a hyperplane.

the generic yield condition for any hyperplane. Also, $(\cos\beta, \sin\beta)$ are clearly the direction cosines of the outward normal to the hyperplane.

In general, the yield conditions for node j of element i can be written concisely as

$$\mathbf{N}_j^{iT}\mathbf{s}_j^i \leq \mathbf{r}_j^i \tag{6.6}$$

where, for the case shown in Figure 6.15,

$$\mathbf{N}_j^{iT} = \begin{bmatrix} \ell_1 & m_1 \\ \ell_2 & m_2 \\ \ell_2 & -m_2 \\ \ell_1 & -m_1 \\ -\ell_1 & -m_1 \\ -\ell_2 & -m_2 \\ -\ell_2 & m_2 \\ -\ell_1 & m_1 \end{bmatrix}, \quad \mathbf{r}_j^i = \begin{bmatrix} r_{j1}^i \\ r_{j2}^i \\ r_{j2}^i \\ r_{j1}^i \\ r_{j1}^i \\ r_{j2}^i \\ r_{j2}^i \\ r_{j1}^i \end{bmatrix}, \quad \mathbf{s}_j^i = \begin{bmatrix} s_1^i \\ s_2^i \end{bmatrix}.$$

We can now collect the relations that apply to both nodes j and k to provide the yield conditions for element i, or

$$\mathbf{N}^{iT}\mathbf{s}^i \leq \mathbf{r}^i \tag{6.7}$$

which has the same form as Equation (6.6) and where, obviously,

$$\mathbf{N}^{iT} = \begin{bmatrix} \ell_1 & m_1 & 0 \\ \ell_2 & m_2 & 0 \\ \ell_2 & -m_2 & 0 \\ \ell_1 & -m_1 & 0 \\ -\ell_1 & -m_1 & 0 \\ -\ell_2 & -m_2 & 0 \\ -\ell_2 & m_2 & 0 \\ -\ell_1 & m_1 & 0 \\ \ell_1 & 0 & m_1 \\ \ell_2 & 0 & m_2 \\ \ell_2 & 0 & -m_2 \\ \ell_1 & 0 & -m_1 \\ -\ell_1 & 0 & -m_1 \\ -\ell_2 & 0 & -m_2 \\ -\ell_2 & 0 & m_2 \\ -\ell_1 & 0 & m_1 \end{bmatrix}, \quad \mathbf{r}^i = \begin{bmatrix} \mathbf{r}_j^i \\ \mathbf{r}_k^i \end{bmatrix}, \quad \mathbf{s}^i = \begin{bmatrix} s_1^i \\ s_2^i \\ s_3^1 \end{bmatrix}.$$

Finally, we can assemble the yield conditions for the whole structure, which can be represented in the indexless form of Equation (6.7) as

$$\mathbf{N}^T\mathbf{s} \leq \mathbf{r} \tag{6.8}$$

where, as is typical in finite element methodology, vector and matrix quantities represent the unassembled contributions of corresponding elemental entities, as concatenated vectors and block diagonal matrices, respectively. In particular, $\mathbf{s}^T = [\mathbf{s}^{1T} \ldots \mathbf{s}^{nT}]$, $\mathbf{r}^T = [\mathbf{r}^{1T} \ldots \mathbf{r}^{nT}]$, and $\mathbf{N} = \text{diag}[\mathbf{N}^i \ldots \mathbf{N}^n]$. In general, if the yield surface at a node has been piecewise linearized into y hyperplanes, then matrix $\mathbf{N}$ is of size $(3n \times 2yn)$ and $\mathbf{r}$ is a $(2yn \times 1)$ vector.

6.4.2 Dual LP Statements of the Discrete Limit Analysis Problem

Within the framework provided by the static-nodal description we can now formulate the discrete form of the static approach to limit analysis as the following LP problem in variables are α and $\mathbf{s}$:

$$\left.\begin{array}{ll} \text{maximize} & \alpha \\ \text{subject to} & \mathbf{B}s - \alpha\mathbf{f} = \mathbf{0}, \\ & \mathbf{N}^T\mathbf{s} \leq \mathbf{r} \end{array}\right\} \tag{6.9}$$

where $\mathbf{0}$ is a null vector of appropriate size, namely $(d \times 1)$.

It is interesting to note that, at this stage, anyone familiar with mathematical programming will, without needing to understand the physical meaning of the quantities involved, be able to immediately formulate the dual LP problem from known duality properties in LP.[6] The dual problem can be encoded as

$$\left.\begin{array}{ll} \text{minimize} & \mathbf{r}^T\mathbf{z} \\ \text{subject to} & \mathbf{B}^T\mathbf{u} - \mathbf{N}\mathbf{z} = \mathbf{0}, \\ & \mathbf{f}^T\mathbf{u} = 1, \\ & \mathbf{z} \geq \mathbf{0} \end{array}\right\} \tag{6.10}$$

where $\mathbf{z}$ and $\mathbf{u}$ are dual variables.

Mechanically speaking, $\mathbf{z}$ can be recognized as a vector of non-negative plastic multiplier rates and $\mathbf{u}$ as a vector of nodal displacement rates. The first constraint set represents compatibility, and the second a normalized external work term. Note that application of the work equation (equating external work rate and internal dissipation) gives

$$\alpha\mathbf{f}^T\mathbf{u} = \mathbf{r}^T\mathbf{z}$$

and because $\mathbf{f}^T\mathbf{u} = 1$, then the objective function of the LP problem [Equation (6.10)], for any admissible solution (possible mechanism), is in fact the kinematically admissible load factor α_k.

Knowledge of LP provides yet another extremely useful property related to these LP problems, namely that the Lagrange multipliers of the optimal solution to the static (kinematic) problem provide us directly with a set of optimal kinematic (static) variables. Most LP solvers, including the one included in the MATLAB optimization toolbox, will report these multipliers as a by-product of the optimization. Specifically, each constraint in Equation (6.9) is associated with a Lagrange multiplier, and at optimality the elements of the reported Lagrange multiplier vector of length $(d + 2yn)$ are ordered in the order of the static constraint set. For instance, the first d multipliers represent the optimal values for the displacement rates $\mathbf{u}$ (conjugate with the static variables $\alpha\mathbf{f}$), and the next $2yn$ multipliers furnish the optimal plastic multiplier rates $\mathbf{z}$ associated with each of the yield hyperplanes. This useful property can be used to obtain, say, kinematic optimal solutions without having to actually solve the kinematic problem [Equation (6.10)].

6.5 A Simple MATLAB Implementation for Static Limit Analysis

In his 1990 preface to the volume on the series of lectures held at the International Centre for Mechanical Sciences, Udine, Lloyd Smith[7] rightly wrote that the practical espousement of mathematical programming approaches in engineering plasticity is largely dependent on the availability of software that can handle practical structures. After so many years since that workshop, one must lament that, while there are some high-quality codes available, not nearly enough has been achieved in that direction. In many respects, that role needs to be fulfilled by our students—the engineers of tomorrow.

The aim of this section is to provide some key ideas as to how a limit analysis software can be simply constructed. While the particular pilot code described is restricted only to plane frames for which pure moment plastic hinges can be formed, it is general enough to be easily extendable. Moreover, even in its present form, it can handle quite large structures, primarily because of the state-of-the-art so-called interior point LP solver it uses.

The coding has been carried out using the well-known MATLAB technical computing language. As indicated earlier, this language may not be as popular with students as Excel, but MATLAB does provide several key advantages, including, for instance, the ability to handle very large data structures. It should also be mentioned that an alternative, increasingly popular, encoding tool is the use of one of the many sophisticated modeling systems available. We need only mention one of them, namely GAMS (an acronym for General Algebraic Modeling System),[8] that is specifically designed to make the construction and solution of large and complex mathematical programming models easy for the nonmathematically oriented user. Its application for solving the limit analysis problem has been illustrated, for example, by Tin-Loi.[9] GAMS, incidentally, is freely accessible over the internet.[10]

The main features of the MATLAB script described here are as follows: a simple design that follows directly the concepts introduced in the previous section; use of an input data text file that facilitates the solution of different problems; writing of a brief output text file that allows for results to be printed and kept; a structure that can easily be extended by students; and, finally, call to a powerful LP solver (LINPROG), available as part of the optimization toolbox of MATLAB, that can handle very large LP problems.

(a) *Data specification and input.* A text file is used for data specification and because it is separate from but has to be incorporated as part of the MATLAB main script, it must be named with the conventional "**.m**" extension (e.g., **LA_in.m**).

The data specification has been purposely kept as simple as possible. It consists of a **Title**, specification of numbers of elements and degrees of freedom, element data (namely orientations of element and plastic moment capacities), element location vectors, and basic load vector.

As an example, consider the model of the simple propped cantilever analyzed previously through Excel. A static-nodal model is shown in Figure 6.17 and the corresponding input file is shown in Figure 6.18. It should be noted that the orientation $j-k$ and θ (Figure 6.13) is given in terms of directed projections. In particular, array **El_dat** contains the signed x projections in column1, the signed y projections in column2, and the plastic moment capacities in column3 (with each row of **El_dat** representing an element). This data structure allows for easy extension of the data specification to include, for example, combined stresses (e.g., by adding plastic axial capacity information in column4).

(b) *Read and check data input.* The main MATLAB script starts off with clearing all variables, reading the input file **LA_in.m**, and performing some checks for size conformity, as in Figure 6.19. Additional checks, including a plotting facility if desired, can be added to the script.

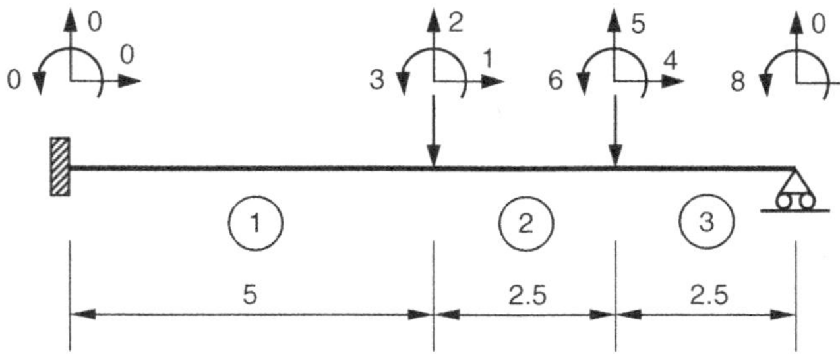

FIGURE 6.17. Model of simple propped cantilever.

```
Title = 'Example 1: simple propped cantilever';

Num_el  = 3;
Num_dof = 8;

El_dat = [ 5    0  10
           2.5  0  10
           2.5  0  10 ];

LV =     [ 0  0  0  1  2  3
           1  2  3  4  5  6
           4  5  6  7  0  8 ];

f = zeros(Num_dof,1);
f(2) = -1;
f(5) = -4;
```

FIGURE 6.18. Input file LA_in.m.

```
% Limit Analysis of Plane Frames
%   Caters only for pure moment hinges
%   Stress resultant s = (n,m1,m2)
%   LP solver LINPROG
% Script : LA.m
% Input  : LA_in.m
% Output : LA.out

% F. Tin-Loi: 22 Nov 92
% Revised:    15 Mar 07

clear all

% Input data (from data file)
% ---------------------------
LA_in

% Input element data : dx dy Mp
% Input location vectors LV
% Input load f

% Check size
% ----------

% Check number of elements
if Num_el ~= size(LV,1)
  error('No. elements incompatible with LV size');
end;

% Check number of dof
if Num_dof ~= max(max(LV))
  error('No. dof incompatible with LV size')
end;
```

FIGURE 6.19. Data input and checking.

(c) *Assemble equilibrium matrix.* The next phase involves the calculation of element equilibrium matrices and the assembly of the structure equilibrium matrix **B**. The code faithfully follows the procedure described in Section 6.1. Location vectors are used for the assembly process and, because MATLAB does not allow zero index values, a simple artifice is used to "assemble" elements corresponding to zero

```
% Assemble equilibrium matrix B
% ----------------------------

% Zero equilibrium matrix B
% Add in a dummy row at dof+1 for freedom = 0
B = zeros(Num_dof+1,3*Num_el);

% Change all 0 in LV to Num_dof+1
LV(LV==0) = Num_dof+1;

% Assemble by looping through all elements
for i=1:Num_el

  % Generate equilibrium sub-matrix b
  L  = sqrt(El_dat(i,1)^2+El_dat(i,2)^2);
  cs = El_dat(i,1)/L;
  sn = El_dat(i,2)/L;
  sL = sn/L; cL = cs/L;
  b = [  cs -sL -sL
         sn  cL  cL
         0   1   0
        -cs  sL  sL
        -sn -cL -cL
         0   0   1  ];
  % Assemble
  ii = i*3-2;
  jj = ii+2;
  B(LV(i,:),ii:jj) = b;

end;

% Delete last "dummy" row to form B
B = B(1:Num_dof,:);
```

FIGURE 6.20. Construction of a structure equilibrium matrix.

location vectors. What has been done is to introduce an additional freedom, assemble zero freedoms into that location, and then finally delete that row from the structure equilibrium matrix. The associated code fragment implementing this is given in Figure 6.20.

(d) *Form LP problem for solution with LINPROG.* The way data are organized for processing by the optimizer largely depends on the particular solver used. In our case, we use LINPROG, a large-scale interior point solver. Because we have restricted ourselves to plastic hinges formed only by pure bending, the yield conditions are simply enforced as upper and lower bounds on the stress variables. As shown in Figure 6.21, results are recovered from the quantities **alpha** (optimal load multiplier), **s_alpha** (vector containing the values of all variables at optimality), and **lambda** (the dual optimal or Lagrange multiplier values).

(e) *Postprocessing.* This part of the script (Figure 6.22) simply extracts the collapse load and optimal stress resultant values from vector **s_alpha**, the dual displacement vector **u** from the Lagrange multiplier output, and also the indices (namely, whether s_2 or s_3 is active) and locations (namely, which elements) of activated hinges.

```
% Set up LP and solve using Matlab LINPROG solver
% ---------------------------------------------

% Find number of variables s and alpha
Num_var = Num_el*3+1;

% Add load vector to last column of B
B(:,Num_var) = -f;

% Objective function coefficients for min (-alpha)
obj(1,Num_var) = -1;

% RHS b
b = zeros(Num_dof,1);

% Set bounds on variables s and alpha
VUB = zeros(Num_var,1);
VUB(:,1) = inf;
for i=1:Num_el
  VUB(i*3-1,1) = El_dat(i,3);
  VUB(i*3,1) = El_dat(i,3);
end;
VLB = - VUB;
VLB(Num_var,1) = 0;  % LB on load factor = 0

% Solve LP
% All constraints are equalities, RHS b = 0
% All variables specified with UB and LB

% Call LINPROG
[s_alpha,alpha,EXITFLAG,OUTPUT,lambda] = ...
    linprog(obj,[],[],B,b,VLB,VUB);
```

FIGURE 6.21. Setting up and solving the LP problem.

```
% Process results
% --------------

% Extract alpha and s
alpha = s_alpha(Num_var,1);
s = s_alpha(1:Num_var-1);

% Lagrange multiplier 'lambda' contains dual values
%   lambda.eqlin ... deflection
%   lambda.lower ... active LB (negative yield)
%   lambda.upper ... active UB (positive yield)

% Deflections (-ve sign, conjugate with -f)
u = -lambda.eqlin;

% Extract yield mode indices
lam_neg = lambda.lower;
lam_pos = lambda.upper;
Index_yield_neg = find(lam_neg > 1e-6)';
Index_yield_pos = find(lam_pos > 1e-6)';

% Find element number and active stress (s2 or s3)
el_neg = ceil(Index_yield_neg./3);
s_neg = Index_yield_neg-3*(el_neg-1);
el_pos = ceil(Index_yield_pos./3);
s_pos = Index_yield_pos-3*(el_pos-1);
```

FIGURE 6.22. Postprocessing the results.

```
% Write output file
% -----------------

fname = 'LA.out';
fid = fopen(fname,'w');
MyDate = date;
fprintf(fid,'Title :     %s\n',Title);
fprintf(fid,'Date  :     %s\n\n',MyDate);
fprintf(fid,'alpha :     %10.6f\n\n',alpha);
fprintf(fid,'Stress resultants s:\n');
for i = 1:Num_el
  fprintf(fid,'  El  %-4d %+6.4e  %+6.4e  %+6.4e\n',...
      i,s(3*i-2:3*i));
end;
fprintf(fid,'\n');
fprintf(fid,'Displacements u:\n');
for i = 1:Num_dof
  fprintf(fid,'  dof %-4d %+6.4e\n',i,u(i));
end;
fprintf(fid,'\n');
fprintf(fid,'Negative hinges (element, s2 or s3):\n');
for i=1:size(el_neg,2)
  fprintf(fid,'  El  %-4d s%-6d\n',el_neg(i),s_neg(i));
end;
fprintf(fid,'Positive hinges (element, s2 or s3):\n');
for i=1:size(el_pos,2)
  fprintf(fid,'  El  %-4d s%-6d\n',el_pos(i),s_pos(i));
end;
fclose(fid);
```

FIGURE 6.23. Output of the results.

(f) *Output results to a file.* Finally, some key results are written to a text file named **LA.out**, as detailed in Figure 6.23. As an example, the output obtained from running actual data specified in Figure 6.18 is shown in Figure 6.24. The student should verify that these results agree with the Excel run.

```
Title :     Example 1: simple propped cantilever
Date  :     15-Mar-2007

alpha :       1.428571

Stress resultants s:
  El  1     +0.0000e+000  +1.0000e+001  +5.7143e+000
  El  2     +0.0000e+000  -5.7143e+000  +1.0000e+001
  El  3     +0.0000e+000  -1.0000e+001  +0.0000e+000

Displacements u:
  dof 1     +0.0000e+000
  dof 2     -1.4286e-001
  dof 3     -2.8571e-002
  dof 4     +0.0000e+000
  dof 5     -2.1429e-001
  dof 6     +2.8571e-002
  dof 7     +0.0000e+000
  dof 8     +8.5714e-002

Negative hinges (element, s2 or s3):
  El  3     s2
Positive hinges (element, s2 or s3):
  El  1     s2
  El  2     s3
```

FIGURE 6.24. Output for simple propped cantilever example.

6.6 A Note on Optimal Plastic Design of Frames

We conclude this chapter with a note concerning the design or synthesis problem, as opposed to the analysis or state problem so far considered. Our treatment of this minimum weight (or volume) problem is not via certain known results such as Foulkes' classical theorems (e.g., Neal[3]) but through a computation-based mathematical programming approach.

We retain all the assumptions made in Section 6.4.1 for the limit analysis problem, except that we assume that bending only causes the formation of plastic hinges in the minimum weight problem. In addition, we make two new and important assumptions. First, it is assumed that a continuous variation of section properties is available, although in practice only a discrete number of sections is available. The implication is that, mathematically, the underlying problem becomes one of continuous optimization instead of the far more challenging discrete (or integer programming) optimization problem. Second, although the variation of typical beam sections is such that their weight per unit length varies nonlinearly with the plastic moment capacity (e.g., in proportion to $M_p^{0.6}$ for beam sections), we assume that the variation is linear. Justification for this is provided for instance by Neal,[3] who indicated that it is unlikely that a large range of values of M_p will need to be considered in the minimum weight problem. We note that there are ways in which both nonlinear variation and combined stresses can be accommodated in the minimum weight algorithm (e.g., Tin-Loi[11]) but these aspects are not discussed here.

The assumption that weight per unit length is of the form $(a + bM_p)$ is a particularly important one, as the total weight W of the structure can now be written as

$$W = \sum_i (a + bM_p)^i L^i = a\sum_i L^i + b\sum_i M_p^i L^i.$$

Clearly, minimizing W can be achieved by minimizing the alternative weight (or objective function for the optimization) w given by

$$w = \sum_i M_p^i L^i. \tag{6.11}$$

The limit analysis problem [Equation (6.3) or (6.9)] can now be modified to provide a formulation for the minimum weight problem; the latter is used if more generality is required, especially if the MATLAB code described in Section 6.5 is used as the basis for the synthesis problem. We should also remember that the load $\mathbf{p} = \alpha\mathbf{f}$ is now known. Since we wish to solve our illustrative example with manually calculated data in Excel, we adopt a modification of Equation (6.3). The optimization problem, which requires minimizing the weight of the frame and identifying in the process plastic moment capacities of all members subject to the usual constraint of equilibrium and yield conformity, can be written as

$$\left.\begin{aligned} &\text{minimize} && \sum_i M_p^i L_i \\ &\text{subject to} && \mathbf{m} = \mathbf{B}_0\mathbf{p} + \mathbf{B}_1\mathbf{x}, \\ & && -\mathbf{M}_p \leq \mathbf{m} \leq \mathbf{M}_p. \end{aligned}\right\} \quad (6.12)$$

Because objective function and constraints are all linear, we obtain an LP problem. It is also worthy of note that, as for the limit analysis problem, a dual LP problem can be written; we will not expand on this but will leave it to interested students who are familiar with mathematical programming to do so.

We conclude this section with a classical simple frame example, considered for instance by Neal[3] in nondimensional form. We present a dimensional version of the same problem in Figure 6.25a. Also shown (Figure 6.25b) are the three assumed redundants used to generate the required moment expressions for the seven critical sections indicated. The beam is assumed to be uniform with a plastic moment capacity of r_1 while both uniform columns have identical moment capacities of r_2.

In the following, we state the LP problem, for which combined equilibrium and yield conformity constraints for each of these critical sections, as detailed previously in Section 6.3, have to be written (Constraint1 to Constraint14). Note also that a pair of constraints, related to positive and negative bending, has to be specified for each check point. We have arbitrarily assumed that positive bending produces tension on the inner face of the frame. Finally, the nonnegativity conditions for r_1 and r_2 have been specified, respectively, as Constraint15 and Constraint16.

The LP problem is as follows:

Objective: $\min w = 12r_1 + 8r_2$
Variables: h, v, m, r_1, r_2

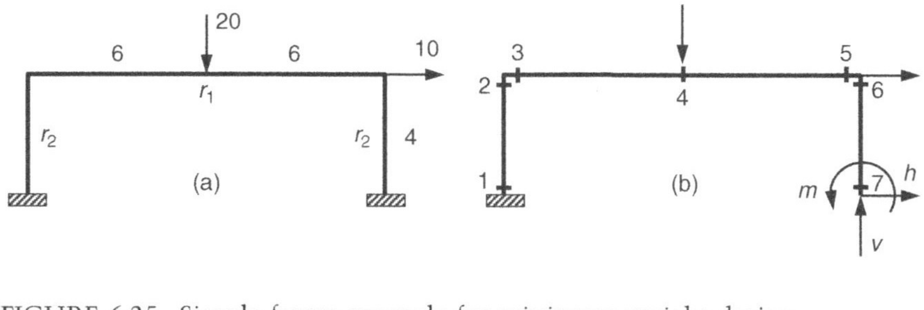

FIGURE 6.25. Simple frame example for minimum weight design.

Constraint1: $12v + m - r_2 \leq 160$
Constraint2: $12v + m + r_2 \geq 160$
Constraint3: $4h + 12v + m - r_2 \leq 120$
Constraint4: $4h + 12v + m + r_2 \geq 120$
Constraint5: $4h + 12v + m - r_1 \leq 120$
Constraint6: $4h + 12v + m + r_1 \geq 120$
Constraint7: $4h + 6v + m - r_1 \leq 0$
Constraint8: $4h + 6v + m + r_1 \geq 0$
Constraint9: $4h + m - r_1 \leq 0$
Constraint10: $4h + m + r_1 \geq 0$
Constraint11: $4h + m - r_2 \leq 0$
Constraint12: $4h + m + r_2 \geq 0$
Constraint13: $m - r_2 \leq 0$
Constraint14: $m + r_2 \geq 0$
Constraint15: $r_1 \geq 0$
Constraint16: $r_2 \geq 0$

The corresponding Excel setup is shown in Figure 6.26. Values of variables and constraints are stored in corresponding cells located in column H. We have also included a table of variable coefficients in

	A	B	C	D	E	F	G	H	I
1	Minimum weight frame								
2									
3	h							0	
4	v							0	
5	m							0	
6	r1							0	
7	r2							0	
8	w							0	
9		h	v	m	r1	r2	RHS		
10	Constraint1		12	1		-1	160	0	
11	Constraint2		12	1		1	160	0	
12	Constraint3	4	12	1		-1	120	0	
13	Constraint4	4	12	1		1	120	0	
14	Constraint5	4	12	1	-1		120	0	
15	Constraint6	4	12	1	1		120	0	
16	Constraint7	4	6	1	-1			0	
17	Constraint8	4	6	1	1			0	
18	Constraint9	4		1	-1			0	
19	Constraint10	4		1	1			0	
20	Constraint11	4		1		-1		0	
21	Constraint12	4		1		1		0	
22	Constraint13			1		-1		0	
23	Constraint14			1		1		0	
24	Constraint15				1			0	
25	Constraint16					1		0	
26									
27									

FIGURE 6.26. Excel sheet for minimum weight problem.

Target Cell (Min)

Cell	Name	Original Value	Final Value
$H$8	w	0	600

Adjustable Cells

Cell	Name	Original Value	Final Value
$H$3	h	0	-10
$H$4	v	0	10
$H$5	m	0	10
$H$6	r1	0	30
$H$7	r2	0	30

Constraints

Cell	Name	Cell Value	Formula	Status	Slack
$H$10	Constraint1	100	$H$10<=$G$10	Not Binding	60
$H$11	Constraint2	160	$H$11>=$G$11	Binding	0
$H$12	Constraint3	60	$H$12<=$G$12	Not Binding	60
$H$13	Constraint4	120	$H$13>=$G$13	Binding	0
$H$14	Constraint5	60	$H$14<=$G$14	Not Binding	60
$H$15	Constraint6	120	$H$15>=$G$15	Binding	0
$H$16	Constraint7	0	$H$16<=$G$16	Binding	0
$H$17	Constraint8	60	$H$17>=$G$17	Not Binding	60
$H$18	Constraint9	-60	$H$18<=$G$18	Not Binding	60
$H$19	Constraint10	0	$H$19>=$G$19	Binding	0
$H$20	Constraint11	-60	$H$20<=$G$20	Not Binding	60
$H$21	Constraint12	0	$H$21>=$G$21	Binding	0
$H$22	Constraint13	-20	$H$22<=$G$22	Not Binding	20
$H$23	Constraint14	40	$H$23>=$G$23	Not Binding	40
$H$24	Constraint15	30	$H$24>=$G$24	Not Binding	30
$H$25	Constraint16	30	$H$25>=$G$25	Not Binding	30

FIGURE 6.27. Solution report for minimum weight problem.

columns B to F to allow for easy data checking and generation of constraint expressions.

Solution of this LP problem using the Excel solver add-in provides the report displayed in Figure 6.27. As indicated, the optimal solution is $w = 600$, $r_1 = r_2 = 30$, $h = -10$, $v = 10$, and $m = 10$. These optimal plastic capacities are exactly as given by Neal.[3]

It is more difficult, however, to extract from Excel the optimal collapse mechanism, which, it must be noted, is not necessarily unique. The "status" of the constraints indicates if they are binding, but a "binding" status does not necessarily indicate if the corresponding hinge is active (i.e., if it contributes to the mechanism). Of course, if a dual solution were available then the nature of yielded sections (whether active or inactive) could be determined.

For this particularly simple frame, results indicate the 2 degrees of freedom mechanism shown in Figure 6.28; the hinge at critical section 1 is inactive. The reason why this type of mechanism is possible is that both beam and columns have the same plastic moment

FIGURE 6.28. Feasible collapse mechanism.

capacities. Neal[3] discussed this aspect in some depth and provided, through Foulkes' geometrical analogue (mapping possible solutions on the $r_1 - r_2$ space), alternative feasible mechanisms.

Bibliography

1. Baker, J. F., Horne, M. R., and Heyman J. (1956). *The steel skeleton, Volume II: Plastic behavior and design.* Cambridge University Press, UK.
2. Massonnet, C. E., and Save, M. A. (1965). *Plastic analysis and design, Volume 1: Beams and frames.* New York: Blaisdell Publishing Co.
3. Neal, B. G. (1963). *The plastic methods of structural analysis.* London: Chapman and Hall.
4. Charnes, A., and Greenberg, H. J. (1951). Plastic collapse and linear programming. *Bulletin of the American Mathematics Society,* **57**, p. 480.
5. Cohn, M. Z., Maier, G. (Eds.) (1979). *Engineering plasticity by mathematical programming.* New York: Pergamon Press.
6. Chvátal, V. (1983). *Linear programming.* New York: Freeman.
7. Lloyd Smith, D. (Ed.) (1990). *Mathematical programming methods in structural plasticity.* New York: CISM Courses and Lectures, Springer Verlag.
8. Brooke, A., Kendrick, D., Meeraus, A., and Raman, R. (1998). *GAMS: A user's guide.* Washington, DC: GAMS Development Corporation.
9. Tin-Loi, F. (1995). Plastic limit analysis of plane frames and grids using GAMS. *Comp. and St.,* **54**, pp. 15–25.
10. Dolan, E. D., Fourer, R., Moré, J. J., and Munson, T. S. (2002). Optimization on the NEOS server. *SIAM News,* **35**, pp. 8–9.
11. Tin-Loi, F. (1990). On the optimal plastic synthesis of frames. *Eng. Optimization,* **16**, pp. 91–108.

CHAPTER 7

Factors Affecting Plastic Collapse

7.1 Introduction

Plastic analysis has been used traditionally to assess the collapse behavior of structures on the basis of yielding of cross sections under proportionally increasing loading. However, there are some circumstances under which the traditional methods of plastic analysis cannot be applied.

For example, there are materials that may not be able to sustain plastic moment throughout the loading history because of lack of ductility in the materials. In this case, it may be necessary to limit the plastic rotation in the plastic hinges instead of allowing for indefinite plastic rotation capacity. In other cases, instead of increasing the loading, it may be more realistic to increase some prescribed displacements proportionally, such as foundation settlements, in order to realize the effect of settlements on failure behavior of the structure.

For structures under increasing temperature, such as those in fires, it is more relevant to calculate the failure temperature, rather than the failure load, for the assessment of the safety of the structure under high temperature. In this case, proportionally increasing prescribed temperatures may be a better way to assess the collapse behavior of the structure.

This chapter discusses the influence of these factors on the failure of structures and describes ways of analyzing the structures.

7.2 Plastic Rotation Capacity

A prominent feature for structures to be able to perform satisfactorily in a plastic fashion is the ability of the structural members to maintain their plastic moment capacities while undergoing plastic deformations. Some materials are so brittle that plastic deformation at rupture is relatively small or even nonexistent. Such materials, for example, glass, are said to be nonductile. However, some materials, such as steel, are ductile and can sustain large plastic deformation before rupture occurs. Ductile materials are particularly suitable for structural design using the plastic method because of their ability to sustain plastic deformation and to maintain plastic moment capacity. Concrete, although brittle, in conjunction with steel as reinforcement can become ductile. However, whether ductile or not, the ductility of all materials is finite. A more realistic criterion for defining failure in the plastic design of structures may be based on assessing the limit of ductility for the materials when the structural members are undergoing plastic deformation. In the following, plastic deformation refers to plastic rotation only.

7.2.1 Ductility of Steel

The plastic rotation capacity of a material is related to the ductility of the material. Traditionally, a simplified approach to imposing ductility requirements for plastic design is to ensure that the ratio of plastic strain to elastic strain must be greater than a certain prescribed value. A typical requirement of ductility for steel is that shown in Figure 2.20 in which the minimum ratio of plastic strain to elastic strain is 6. This assumption is based on the traditional practice for the plastic design of simple and low-rise structures where the number of plastic hinges is small before collapse occurs. For large rigid-jointed and highly redundant steel structures, some yielded sections may undergo large plastic rotations as the number of plastic hinges grows. In this case, it is more realistic to check the actual plastic rotation capacity against the maximum available plastic rotation, termed "plastic rotation demand," in plastic hinges. The serviceability requirement regarding plastic rotation for design may be written as

$$\theta_d \leq \theta_h \tag{7.1}$$

where θ_d is plastic rotation demand and θ_h is plastic rotation capacity as shown in Figure 7.1.

Work to quantify the plastic rotation capacity of steel members has been carried out previously by a number of researchers.[1,2] An expression given by Ziemian and colleagues[3] for the plastic rotation capacity of steel I sections, modified from an expression originally developed by Lay and Galambos,[1] is adopted and given here:

$$\theta_h = 2.84\varepsilon_y(s-1)\left[\left(\frac{bt}{dw}\right)\left(\frac{A_w}{A_f}\right)\right]^{1/4}\left(1+\frac{V_1}{V_2}\right)\left(\frac{h_w}{2h_c}\right) \tag{7.2}$$

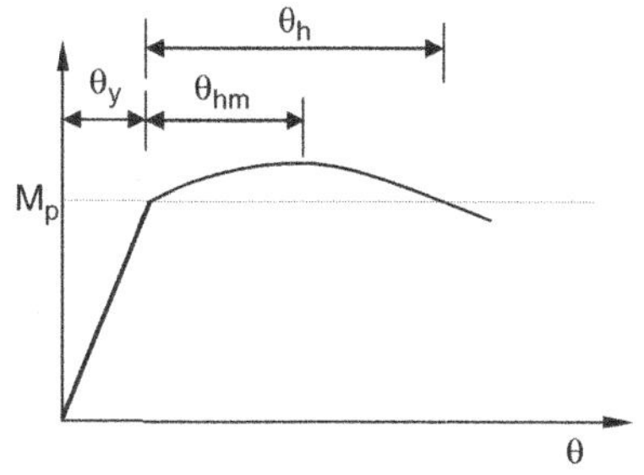

FIGURE 7.1. Plastic rotation.

where ε_y is yield strain, s is ratio of strain-hardening strain (ε_s) to yield strain $(\varepsilon_y) \approx 6$ for steel (see Figure 2.1), b is flange width, t is flange thickness, d is section depth, w is web thickness, V_1/V_2 (≤ 1.0) is ratio of the magnitudes of the shears on both sides of the plastic hinge, h_w is distance between centers of flanges, and h_c is distance between compression flange and the plastic neutral axis. For yielding due to pure bending, $h_w = 2h_c$. For yielding due to axial force-bending interaction, the plastic rotation capacity is reduced by the ratio $h_w/2h_c$ (≤ 1.0) as suggested by Kemp.[2]

In order to ensure that the plastic rotation capacity takes into account the flange or web buckling, an additional requirement, also given by Kemp,[2] needs to be checked:

$$\theta_d/\theta_y \leq R_m (= \theta_{hm}/\theta_y) \tag{7.3}$$

where θ_y is maximum elastic rotation at the plastic hinge, θ_{hm} is inelastic rotation at which ultimate bending moment occurs, and R_m is given by

$$R_m = l\left(2s - 1 + e\frac{l}{1-l}\right)\frac{h_w}{2h_c} \tag{7.4}$$

where l is the plastic proportion length based on flange, web, or lateral buckling of the section and e is ratio of elastic to strain-hardening modulus, approximately equal to 50 for steel. The plastic proportion length l is defined as the ratio of the length of the plastic region to the half-span length. Typically, $R_m = 3$ should be achieved if $l > 0.115$. To ensure that local buckling would not occur during plastic bending, most design codes specify maximum values for the slenderness (b/t) of the flange and the web. In the following, a method to relate the slenderness to R_m is given.

Based on Kemp's work, it was found that within all practical ranges of steel I sections, the flange is likely to buckle first. Therefore, it is assumed in the present analysis that l is based on flange buckling only. A complete theoretical estimation of l can be found in Kemp.[2]

The onset of flange buckling due to bending is related to the plastic proportion length $l = l_f$ by

$$\left(\frac{b}{t_f}\right)_b = \sqrt{\frac{2}{1.5\varepsilon_b - C_f(\pi t_f / l_f L)^2}} \tag{7.5}$$

where $\left(\frac{b}{t_f}\right)_b$ is the ratio of flange width-to-thickness at buckling; $C_f = 0.5$ for no web or warping restraint, or $= 1.0$ with web and warping restraint; and L is length of maximum moment to point of contraflexure in the beam; for a simply supported beam, L is half of the span length;

$$\varepsilon_b = \varepsilon_{yf}\left(s + 0.5e\frac{l_f}{1 - l_f}\right) \tag{7.6}$$

ε_{yf} = yield strain, approximately equal to 0.00125 for a yield stress of 250 MPa.

Using Equations (7.5) and (7.6), l_f can be solved iteratively and R_m calculated according to Equation (7.4). The use of plastic rotation capacity in plastic analysis is demonstrated in Section 7.3.

7.2.2 Plastic Rotation Demand

Plastic rotation starts when a section reaches its plastic moment. An example of plastic rotation is shown in Figure 7.2a where the plastic hinge of a member attached to a fixed support undergoes a plastic rotation θ_{d1}. For an originally straight member with a plastic hinge within its length such as that shown in Figure 7.2b, the amount of plastic rotation is the sum of θ_{d2} and θ_{d3}.

There are two ways to calculate the plastic rotation in plastic analysis according to whether the "condensation method" or the "extra freedom method" (Section 1.12) is used to model the plastic hinge. When using the condensation method, the plastic hinge is modeled implicitly in the element stiffness matrix described in Section 4.4.1 and the plastic rotation is calculated separately in

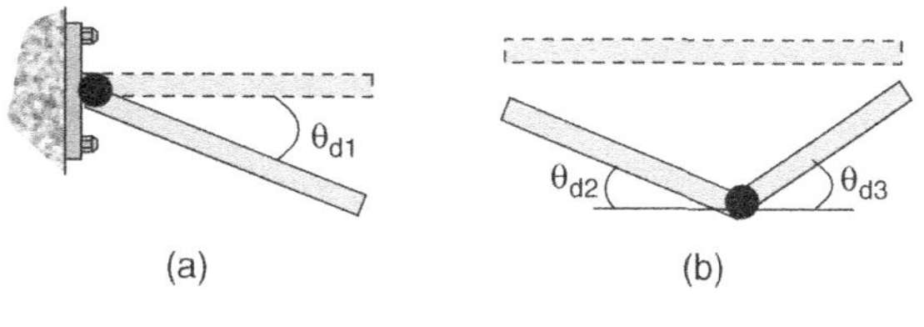

FIGURE 7.2. Plastic rotation.

Equation (3.25). When using the "extra freedom method," the plastic hinge is modeled explicitly in the element stiffness matrix and the plastic rotation is obtained directly from the displacement vector in the solution of the structure equilibrium equation.

Example 7.1 A fixed-end beam ABC shown in Figure 7.3 is subjected to an applied load P at B. Calculate the plastic rotation demand of the plastic hinges at collapse. $E = 2 \times 10^8$ kN/m^2, $A = 0.00764$ m^2, $I = 0.000216$ m^4, $M_p = 324$ kNm.

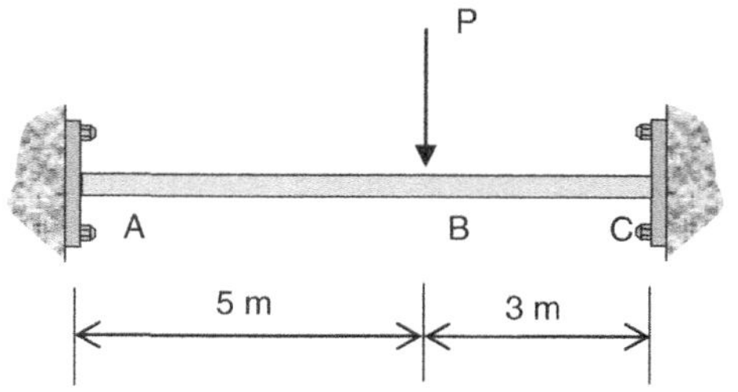

FIGURE 7.3. Fixed-end beam for Example 7.1.

Solution. For this structure, three plastic hinges occur in the following order: at sections C, B, and A. The values of plastic rotations at different stages of calculation are shown in Table 7.1.

At analysis stage 1, the first hinge occurs at C at a load of P = 276.48 kN. At analysis stage 2, the second plastic hinge occurs at B at an increment of P = 58.18 kN. During this stage, section C undergoes a plastic rotation of 3.16×10^{-3} radians. At analysis stage 3, the third plastic hinge occurs at A at an increment of P = 10.91 kN. During this stage, section B undergoes a plastic rotation equal

TABLE 7.1
Results for Example 7.1

Stage 1	*Increment in P*	*Increment in plastic rotation (radian)*		
		At C	*At B*	
		θ_c	θ_{B1}	θ_{B2}
1	276.48			
2	58.18	3.16×10^{-3}		
3	10.91	3.51×10^{-3}	3.16×10^{-3}	3.51×10^{-3}
Total	345.57	6.67×10^{-3}	6.67×10^{-3}	

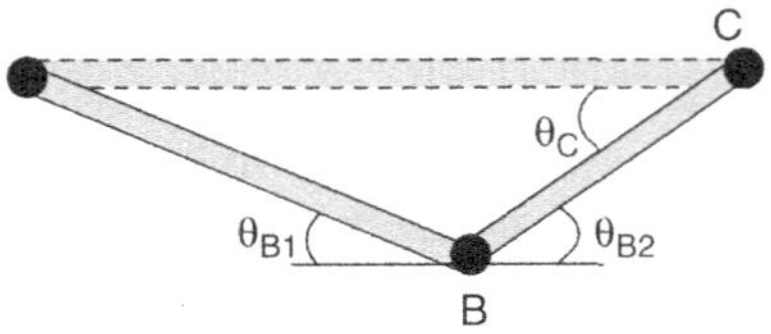

FIGURE 7.4. Plastic rotations at plastic hinges.

to the sum of rotations at the ends of members AB and BC while section C undergoes a further plastic rotation of 3.51×10^{-3} radians. The angles of plastic rotation at B and C are shown in Figure 7.4.

At B, $\theta_{B1} = 3.16 \times 10^{-3}$ radians and $\theta_{B2} = 3.51 \times 10^{-3}$ radians, giving a total plastic rotation of 6.67×10^{-3} radians. When using the extra freedom method, θ_c, θ_{B1}, and θ_{B2} are all obtained from the solution of the structure equilibrium equation.

When using the condensation method, during stage 3 calculation the plastic hinge at B is assumed to occur either in member AB or in member BC in order for the stiffness matrix to be modified. Suppose that the plastic hinge at B occurs in member AB. The incremental solution to the equilibrium equation of the structure gives the vertical deflection of 0.01053 m and an elastic rotation of 3.51×10^{-3} radians at B in BC. Separate calculations give values of plastic rotations of 6.67×10^{-3} radians at B in member AB and 3.51×10^{-3} radians at C in member BC.

7.3 Effect of Settlement

Traditionally, structural design against foundation failure is usually based on the allowable bearing capacity of the supporting soil or the integrity of the foundation members such as rafts and piles. The allowable bearing capacity is governed by the consideration of individual foundation settlements, which are compared with allowable limits stipulated in design codes and building regulations. The magnitudes of the actual settlements are usually calculated on the basis of the magnitudes of applied forces obtained independently from structural analysis. It is well known that settlements may induce member forces in addition to those due to design loads. However, the coupling effects between induced member forces in the superstructure and settlements are rarely investigated by design engineers, although some research has been done in this area, which is typically considered a soil–structure interaction problem. A soil–structure interaction problem can be cast as one using a simple model consisting of a single foundation mat resting on soil or using a complex model where the superstructure, foundation, and soil medium are treated as a continuous system. While the simple model does not reflect the effects of settlement on the behavior of the superstructure above the foundations, the complex model involves large amounts of information and requires sophisticated computational techniques, such as the finite element method, which may not be entirely suitable for routine use by practicing engineers.

A common approach to the control of differential settlements in foundations is to limit the damage caused by angular distortion, which is measured as the ratio of the difference of settlements between adjacent foundations to the distance between them. The maximum allowable angular distortion for buildings varies according to the damage limit requirements in the buildings, but a nominal value of 1/300[4] seems reasonable for limiting the damage to general building services.

The use of angular distortion for controlling foundation settlement is, in some cases, considered overly simplified when assessing structural damage. A more rational approach to the design of foundations may be to evaluate the actual settlement effects on the strength and serviceability requirements of the structure. The plastic design concept can be brought in to check the strength requirements of the structure. In this approach, the advantage of the inelastic properties of the constructional materials, particularly steel, can be taken by making use of the ability of the materials to accommodate deformation as a consequence of excessive settlement. The following introduces such an approach using serviceability requirements to limit the amount of plastic rotations due to foundation settlement according to Equations (7.1) and (7.3). The method takes into account simultaneous deformations in the foundation, including vertical, lateral, and rotational movements. This method can also be used to monitor effects of foundation settlement in existing structures.

7.3.1 Plastic Analysis Due to Settlement Effects

In a settlement problem, attention is usually centered on the movement of the foundation in the vertical direction, although, in reality, lateral and rotational movements also exist. These lateral and rotational movements may be because of eccentric and inclined loads acting on the foundation, uneven movement in the soil, or other factors. This is illustrated in Figure 7.5 where the foundation is displaced from point A to point B.

In addition to the vertical movement Δ, a lateral movement δ and a rotational movement β may also exist. Under this situation, forces are induced in the member associated with these movements and are subsequently redistributed to other parts of the structure.

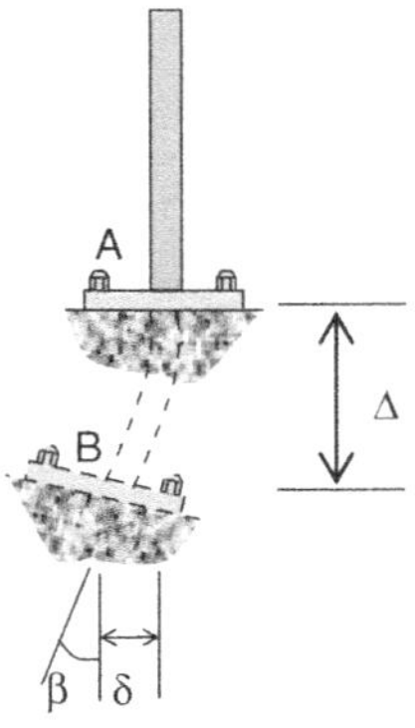

FIGURE 7.5. Foundation movements.

The magnitudes of the induced forces increase as the magnitudes of Δ, δ, and β increase. When the foundation movement continues, the forces in some members become so great that yielding occurs in the form of plastic hinges. At this stage, some limiting values of Δ, δ, and β may be reached before the serviceability requirements of the structure are violated. The desire to calculate these limiting values leads to the concept of "ultimate settlement factor", in contrast to the "collapse load factor" in a plastic analysis.

Consider a propped cantilever beam ABC subjected to a vertical load P acting at midspan as shown in Figure 7.6. In order to investigate the effects of settlement at the support on the behavior of the structure, a prescribed displacement δ at A is assumed. A settlement factor, multiplied to δ and increasing until some serviceability requirements of the structure are breached, can be used to assess the settlement effects while the externally applied load P remains constant. The settlement factor, at its limit, is herein termed the "ultimate settlement factor", α_u, at which the structure is no longer serviceable.

If the settlement δ of the beam in Figure 7.6 keeps increasing, the section at C will eventually become yielded. Indeed, when the settlement δ reaches a value at which section C becomes yielded, any

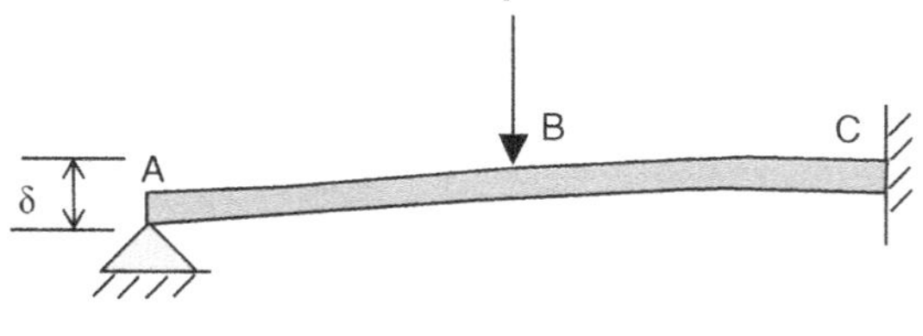

FIGURE 7.6. Settlement of a propped cantilever beam.

further increase in settlement at A will only induce plastic rotation at C, while the structure will not form a collapse mechanism when the applied load is increased until collapse. In this case, the rotation capacity of the plastic hinge described in Section 7.2, which can be used as the serviceability requirement of the structure, is checked against the plastic rotation demand due to foundation settlement.

It must be emphasized that for some structures there may be various settlement modes for different settlement scenarios. In using this method, a number of possible settlement modes can be assumed and, by carrying out plastic analysis and comparing α_u, the worst settlement scenario can be predicted. The method can also be used to monitor the safety level of a structure where known settlements occur.

7.3.2 Modified End Actions Due to Settlements

Forces induced in members as a consequence of support settlement are often treated as internal loads and calculated as a set of fixed-end forces in the stiffness method of analysis. In elastoplastic analysis where the direct method (Section 4.4.1) is used, the fixed-end forces have to be modified in order to take into account the presence of plastic hinges in the members. The forces derived from these modified fixed-end forces are termed "modified end actions" (MEA). A general description of the method to calculate MEA has been given in Section 3.8.

The calculation of MEA due to settlement for members with yielding due to pure bending is straightforward. For example, a member with one hinge at one end can be treated as a propped cantilever with one end displaced by settlements. The general expression of MEA vector is given by Equation (3.66) as

$$\{\Delta P\} = [K_P]\{\Delta d\} + \{\Delta P^M\} \tag{7.7}$$

where the MEA vector due to settlement is given by

$$\{\Delta P^M\} = [K_P]\{D^0\} \tag{7.8}$$

in which the settlement displacement vector $\{D^0\}$ contains the general settlement values (Δ, δ, and β shown in Figure 7.5) associated with the member. For a member with settlement at end j, the settlement displacement vector is written explicitly as

$$\{D^0\} = \{0 \quad 0 \quad 0 \quad \Delta \quad \delta \quad \beta\}^t \tag{7.9}$$

Because $[K^P]$ varies with both the state of plasticity and the yield criterion adopted for the member, the form of MEA also varies. For example, if a horizontal beam of length L with a plastic hinge due to yielding by pure bending at end j is subjected to a vertical settlement δ also at end j, the MEA vector, using Equation (3.52) for $[K_P]$, can be calculated as

$$\{\Delta P^M\} = \begin{Bmatrix} 0 \\ -\dfrac{3EI}{L^3}\delta \\ -\dfrac{3EI}{L^2}\delta \\ 0 \\ \dfrac{3EI}{L^3}\delta \\ 0 \end{Bmatrix} \tag{7.10}$$

Example 7.2 A two-story, two-bay frame with fixed supports A, B, and C is subjected to design loads shown in Figure 7.7.

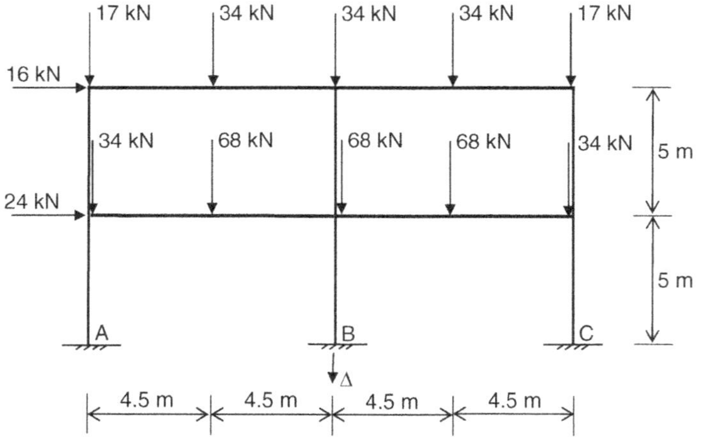

FIGURE 7.7. Example with settlement effect.

All beams at the top level are 200UB25.4, at the lower level are 310UB40.4, and all columns are 150UC37.2. It is predicted that settlement would occur only vertically at the support at B. The initial forces in the members due to the design loads are first calculated by elastoplastic analysis (it happens that no plastic hinge occurs under the design loads). An elastoplastic analysis is then carried out by imposing an increasing vertical settlement Δ at B. Yielding by pure bending is assumed in the analysis, and the rotation capacity at plastic hinges is governed by Equation (7.1) only.

Solution. The following assumptions are made in calculating the rotation capacities at the plastic hinges: $s = 6$, $\varepsilon_y = 0.00125$, $h_w = 2h_c$, $V1/V2 = 1$. The properties of the members are given in Table 7.2.

TABLE 7.2
Properties of members in Example 7.2

Member	M_p *(kNm)*	θ_h *(radian)*
150UC37.2	72	0.042
200UB25.4	61	0.032
310UB40.4	146	0.032

If the plastic rotation capacity requirement is not considered, the formation of the plastic hinges follows the sequence shown in Figure 7.8 and the amounts of plastic rotations of the plastic hinges at increasing Δ are shown in Figure 7.9. The analysis shows that after formation of the 11th plastic hinge, the structure offers no resistance to further foundation settlement at B.

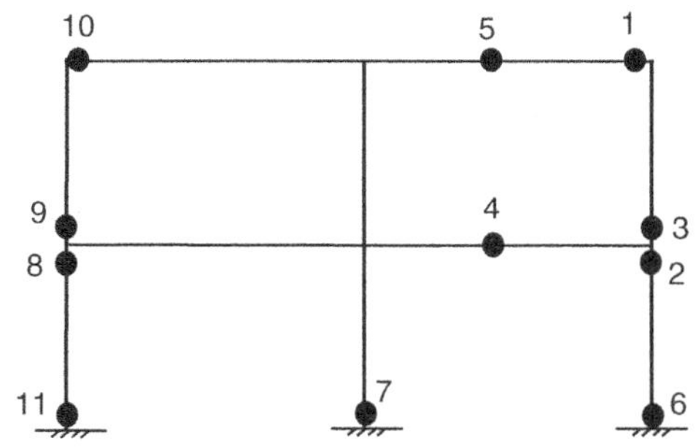

FIGURE 7.8. Plastic hinge formation sequence.

If plastic rotation capacity requirement is considered, it can be seen from Figure 7.9 that hinge 1 reaches its rotation capacity of 0.032 radians first at a total settlement $\Delta = 206$ mm. This is also equal to the ultimate settlement factor ($\alpha_u = \Delta$) at which only the first three plastic hinges occur before the structure is considered no longer serviceable as the rotation capacity of hinge 1 has been reached.

The same frame is again analyzed but the predicted settlements at B are as given in Figure 7.5 in which $\Delta = 10$ mm, $\delta = 2$ mm, and $\beta = 0.002$ radians. This set of settlements is increased proportionally by a common factor α_u. The axial-bending interaction is assumed for the yield criterion.

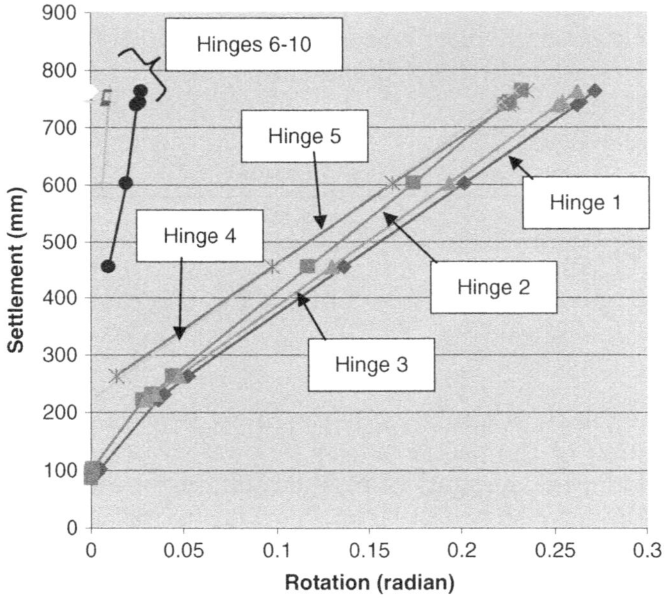

FIGURE 7.9. Plastic hinge rotations.

Results show that the rotation capacity of the plastic hinge at the right hand end of the top beam (hinge 1 in Figure 7.8) is reached with an ultimate settlement factor $\alpha_u = 20.4$. This is equivalent to $\Delta = 204$ mm, $\delta = 40.8$ mm, and $\beta = 0.04$ radians.

7.4 Effect of High Temperature

Thermal loading as a consequence of temperature rise, usually treated as internal loads in structural analysis using the stiffness method, induces a set of self-equilibrating forces in the structure. With a moderate temperature rise, internal loads are calculated in terms of the thermal expansion of the structural materials. A general method to include the effects of high temperature in elastic analysis has been described in Section 1.13. For completeness of this chapter, some of the equations used in Section 1.13 are used here again.

According to the theory of plasticity, the self-equilibrating forces induced by thermal expansion have no effect on the plastic collapse loads that are obtained in a plastic analysis by increasing the applied loads proportionally until collapse. Therefore, traditional plastic analysis with monotonically increasing loading may not be able to reflect the detrimental effect of rising temperature in a structure.

At very high temperatures, the member forces induced by thermal loading are always associated with changes of the material properties of the member, which need to be considered in analysis. For structural design of structures at high temperatures, the critical temperature at which the structure fails in situations such as fire is often required. Indeed, the performance of structures in fire is an important issue in structural engineering, and analytical techniques are required to deal with structures under severe fire conditions. Sophisticated methods, such as the finite element method, have been developed for analyzing structures in fire (e.g., see Najjar and Burgess[5] and Schleich *et al.*[6]).

In the following, a simple approach to the assessment of plastic collapse of steel structures in fire is introduced. The method can be used in plastic analysis where the critical temperature of the structure, rather than the collapse load, is calculated.

7.4.1 Structures Subject to Uniform Temperature

For a structure under a set of loads P applied at location i ($i = 1, 2, 3, \ldots$) with corresponding deflections δ and a uniform temperature T in member j ($j = 1, 2, 3, \ldots$) undergoing plastic rotation, the virtual work equation in plastic analysis using the upper bound approach for the calculation of the load factor α applied uniformly to the loads at collapse is given by

$$\alpha \sum (P\delta)_i = \sum (M_{prT}\theta)_j \tag{7.11}$$

where the reduced plastic moment M_{prT} at temperature T is given as a function $f(\ )$ by

$$M_{prT} = f(M_{pT}, \beta_N) \tag{7.12}$$

$$\frac{M_{pT}}{M_{p20}} = \frac{f_{yT}}{f_{y20}} = \phi_{yT} \tag{7.13}$$

where β_N is the ratio of the axial force to the squash load and M_{pT} is the plastic moment capacity at temperature T. Equation (7.12) is used for sections subjected to the bending-axial interaction described in Chapter 3, and Equation (7.13) represents the yield stress deterioration rate ϕy_T at rising temperature. The yield stress at $T = 20°\text{C}$ is given as f_{y20}. The yield stress deterioration rate of steel stipulated in the Australian code is given by

$$\phi_{yT} = \frac{905 - T}{690} \qquad \text{for } T \geq 215°\text{C}$$

$$\phi_{yT} = 1 \qquad \text{for } T < 215°\text{C} \tag{7.14}$$

It should be noted that Equation (7.11) is similar to Equation (5.1) except that in the present formulation the plastic moment capacity is temperature dependent. For any member where thermal restraint induces internal forces due to thermal expansion, the axial thermal load is calculated by

$$P_T = E_T A \alpha_t (T - 20) \tag{7.15}$$

where $E_T =$ modulus of elasticity at temperature T and α_t is the coefficient of linear thermal expansion. Equation (7.15a) can be rewritten as

$$P_T = E_{20} A \alpha_t \, \phi_{ET} (T - 20) \tag{7.16}$$

where $\phi_{ET} = \dfrac{E_T}{E_{20}}$ is the deterioration rate of the modulus of elasticity at temperature T given by Equations (1.53a) or (1.53b). The variation of P_T is shown in Figure 1.37.

A critical temperature analysis for a structure can be carried out by setting $\alpha = 1$ in Equation (7.11) in which T becomes the only variable. By solving Equation (7.11), the critical temperature can be obtained. It must be emphasized that, contrary to a plastic analysis for structures at room temperature, the static loads in a critical temperature analysis are to remain constant.

Example 7.3 Determine the critical temperature of the propped cantilever beam made of the steel I section shown in Figure 7.10 when (a) the simply supported end is free to expand and (b) the simple supported end is restrained from expanding. For the beam, $M_p = 324$ kNm, $N_p = 820$ kN, $\alpha_t = 0.000012/°C$, $E = 2 \times 10^8$ kN/m^2, $A = 0.0005$ m^2.

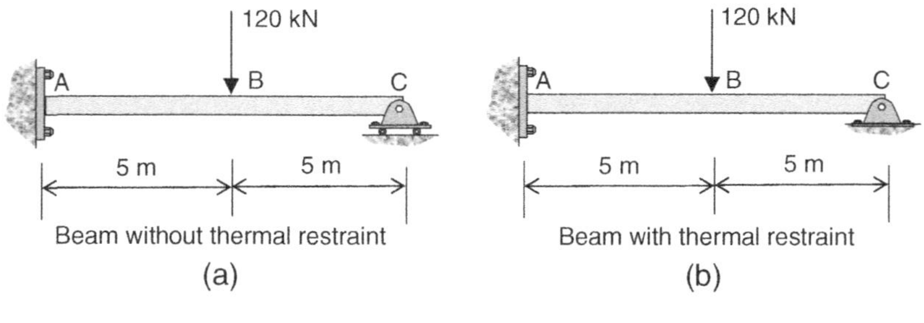

FIGURE 7.10. Beam failure by rising temperature.

Solution. According to the mechanism method, the virtual work equation is

$$M_p(3\theta) = 120(5\theta),$$

therefore

$$M_p = 200 = M_{prT}. \tag{7.17}$$

(a) Because the beam is not thermally restrained,

$$M_{prT} = M_{pT} = 324\left(\frac{905 - T_c}{690}\right)$$

Hence, from Equation (7.17), the critical temperature can be obtained: $T_c = 479°C$.

(b) Because of thermal restraint, the axial force in the beam at critical temperature T_c is

$$N = E_T \times 0.0005 \times 0.000012 \times (T_c - 20) = E_T \times 6 \times 10^{-9}(T_c - 20)$$

Hence,

$$M_{prT} = M_{pT} = 324\left(\frac{905 - T_c}{690}\right) \times 1.18\left(1 - \frac{N}{820\left(\frac{905 - T_c}{690}\right)}\right)$$

Solving Equation (7.17) using, for example, Microsoft Excel gives $T_c = 296°C$.

The aforementioned example shows that the strength of a steel member is reduced greatly by the axial-bending interaction generated by thermal restraints at the supports. For more complex structures, the formation of plastic hinges may help relieve the thermal restraints between members. Prior to collapse, members in the structure may be completely free of thermal restraint as a result of the formation of plastic hinges, which transform the structure into a statically determinate one. The following example shows that prior to collapse at the collapse temperature, the thermal loading induces no forces in the members.

Example 7.4 Determine the critical temperature of the steel portal frame subject to rising uniform temperature shown in Figure 7.11 when $\alpha = 1$. Assume failure by pure bending only. All members are I sections with $M_p = 558$ kNm and $N_p = 2800$ kN.

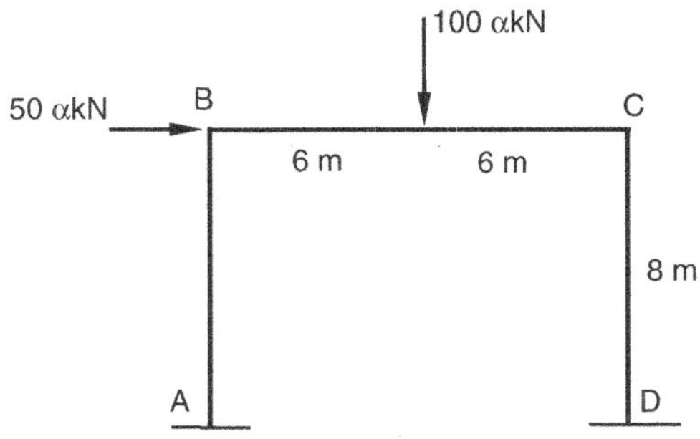

FIGURE 7.11. Portal frame subject to rising uniform temperature.

Solution. The failure mode is a combined beam-sway mechanism shown in Figure 7.12. The critical load factor is $\alpha = \frac{3M_p}{500}$. For $\alpha = 1$ at critical temperature,

$$M_p = \frac{500}{3} = 558\left(\frac{905 - T_c}{690}\right)$$

Hence, $T_c = 699°C$.

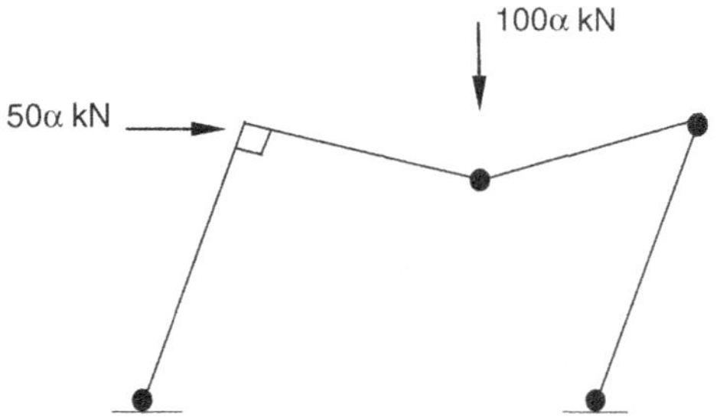

FIGURE 7.12. Combined beam-sway collapse mechanism.

For this example, it may be useful to examine the effect of the axial–force interaction. If the plastic moment capacity of the members is denoted as M_{pT}, the axial forces in the members at collapse are

$$\text{AB: } N_T = \frac{4M_{PT}}{15};\ \text{BC: } N_T = \frac{M_{PT}}{4};\ \text{CD: } N_T = \frac{M_{PT}}{3}.$$

The axial force ratios of the members are

$$\text{AB: } \frac{N_T}{N_{PT}} = \frac{4M_{PT}/15}{N_{PT}} = \frac{4M_P}{15N_P} = 0.05 < 0.15;$$

$$\text{BC: } \frac{N_T}{N_{PT}} = \frac{M_{PT}/4}{N_{PT}} = \frac{M_P}{4N_P} = 0.05 < 0.15;$$

$$\text{BC: } \frac{N_T}{N_{PT}} = \frac{M_{PT}/3}{N_{PT}} = \frac{M_P}{3N_P} = 0.067 < 0.15.$$

Therefore, the members are not subjected to axial-bending effects. Axial force as a consequence of thermal expansion in any member is zero at collapse when the members are free to expand.

The two examples just given are simple enough to enable the critical temperature to be calculated directly. For complex structures, elastoplastic analysis may be used and a direct evaluation of the critical temperature may be difficult. The next section introduces an indirect method to calculate the critical temperature of structures subject to uniform temperature rise.

7.4.2 Critical Temperature Evaluation in Elastoplastic Analysis

This is a trial-and-error method that makes use of the fact that a unique temperature rise in the members corresponds to a unique collapse load factor for the structure. For a uniform temperature rise in the members, the collapse load is independent of the modulus of elasticity. Therefore, variation of the collapse load factor α_c is directly related to the yield strength of the member, which is a function of the temperature. Thus, the relationship between the collapse load factor and the temperature in the structure varies in the same way that the yield strength varies with the temperature. If the collapse load factor–temperature relationship is plotted as in Figure 7.13, the critical temperature T_c of the structure at failure corresponds to a collapse load factor α_c equal to 1. T_c can be obtained by interpolation as shown in Figure 7.13. For a linear relationship between yield strength and temperature, the α_c–temperature relationship is also linear and only two collapse analyses are required to establish the critical temperature by interpolation. To calculate the critical temperature for structures with a nonuniform temperature rise in the members, refer to Wong.[7]

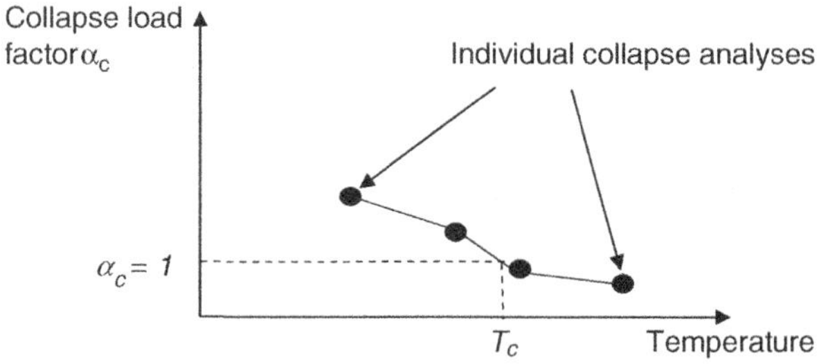

FIGURE 7.13. Collapse load factor–temperature relationship.

Example 7.5 A steel frame made of members with the same size is subjected to design loads shown in Figure 7.14. Determine the critical temperature of the members at which the structure fails. Assume that all members are subjected to uniform temperature rise. For all members, $M_p = 304$ kNm, $E = 2 \times 10^8$ kN/m^2, $A = 0.00689$ m^2, $I = 188 \times 10^{-6}$ m^2.

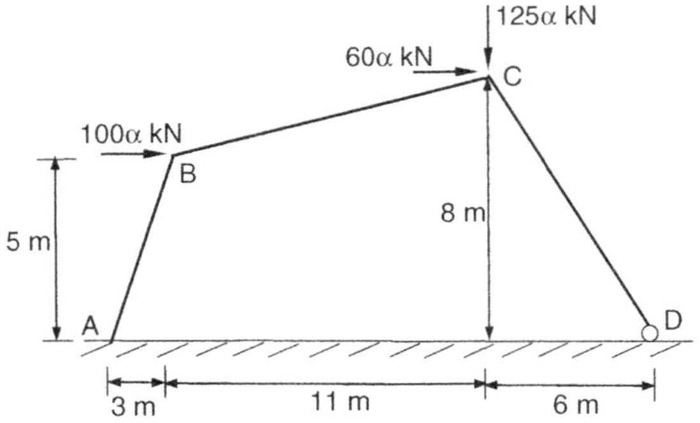

FIGURE 7.14. Steel frame with rising temperature.

Solution. The yield strength deterioration rate of steel with temperature according to Equation (7.14) is used. From elastoplastic analysis, the collapse load factor α_c at room temperature is found to be 2.742 with three plastic hinges formed in the structure. This collapse load factor does not change until the temperature in the structure reaches 215°C when the yield strength of steel starts to deteriorate. At each chosen temperature, the values of M_p and E are calculated and the corresponding collapse load factor is found from the elastoplastic analysis. The following results are obtained: at 300°C, $\alpha_c = 2.404$; at 500°C, $\alpha_c = 1.609$; at 700°C, $\alpha_c = 0.814$. A plot of the α_c–temperature relationship is shown in Figure 7.15. A simple linear interpolation calculation gives $T_c = 653$°C when $\alpha_c = 1$.

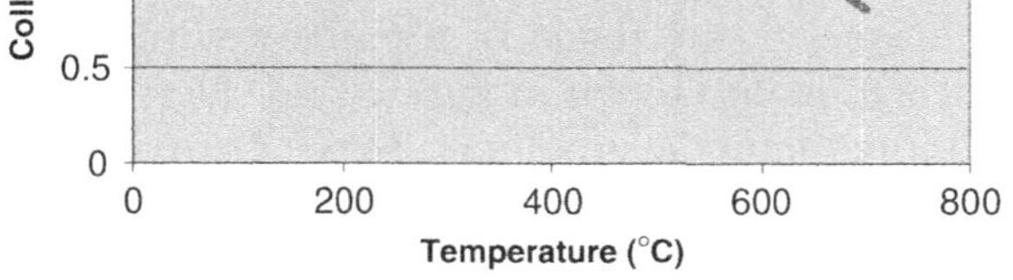

FIGURE 7.15. α_c–temperature relationship.

7.5 Second-Order Effects

Second-order effects include both the effect of nodal displacements (called the P-Δ effect) and the effect of crookedness of the member (called the P-δ effect). The P-Δ effect induces an additional bending moment in a member as a result of the eccentric load acting on the member. The eccentric load occurs due to the relative nodal displacements at the ends of the member. An example is shown in Figure 7.16 in which an additional bending moment equal to P × Δ is induced due to the sway of the frame. The P-δ effect induces an additional bending moment in a member as a result of the combined actions of the axial force acting along the originally straight chord of the crooked member. The crookedness is usually caused by member bending.

Second-order (geometrical nonlinear) analysis can be carried out for structures, taking into account both P-Δ and P-δ effects. In such an analysis, a common load factor is applied to the imposed loads so that by varying the value of the load factor, buckling failure of the structure occurs. At failure due to buckling of the structure, the common buckling load factor is denoted by λ_c. Most commercial structural analysis programs for frames can perform analysis for λ_c. As material yielding is not considered in buckling analysis, the actual failure load factor, λ_a, of the structure is usually lower than λ_c taking into account the effect of material yielding.

A simplified and approximate way to calculate the actual failure load is based on the use of the Merchant–Rankine formula, which takes into account both material and geometrical nonlinear effects. The Merchant–Rankine formula takes the form of

$$\frac{1}{\lambda_a} = \frac{1}{\lambda_p} + \frac{1}{\lambda_c} \tag{7.18}$$

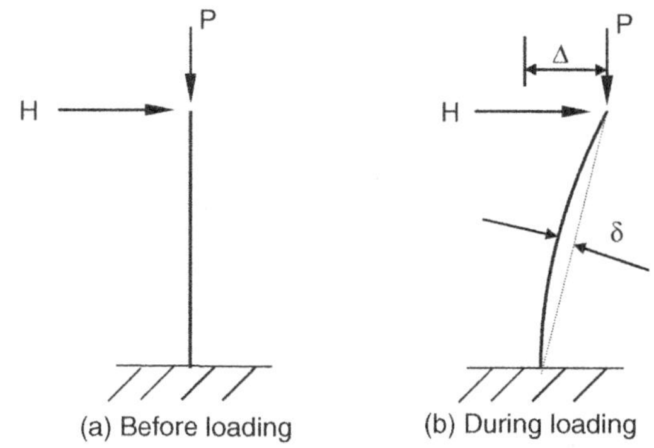

FIGURE 7.16. Second-order effects.

where λ_p is the plastic collapse load factor.

Experiments show that a more accurate estimate of λ_a can be achieved by modifying Equation (7.18) as

$$\frac{1}{\lambda_a} = \frac{0.9}{\lambda_p} + \frac{1}{\lambda_c} \tag{7.19}$$

Hence,

$$\lambda_p = \lambda_a \left(\frac{0.9}{1 - \dfrac{\lambda_a}{\lambda_c}} \right) \tag{7.20}$$

Because practical design requires that the actual failure load factor is equal to the design load factor so that $\lambda_a = 1.0$, a moment amplification factor δ_p can then be defined as

$$\delta_p = \frac{0.9}{1 - \dfrac{1}{\lambda_c}} \tag{7.21}$$

Equation (7.21) is used in design codes to amplify the required plastic moment capacity for plastic design when the upper bound approach is used. Using Equation (7.21), the actual failure load factor can be obtained from modifying the plastic collapse load factor by

$$\lambda_a = \frac{\lambda_p}{\delta_p} \tag{7.22}$$

As seen from the aforementioned derivation, the bending moments in plastic analysis, amplified by δ_p due to both P-δ and P-Δ effects, are equivalent to reducing the plastic collapse load factor by δ_p. In the Australian steel design code (AS4100-1998), use of the moment amplification factor is in accordance with the following rules:

a. For $\lambda_c \geq 10$, second order effects can be ignored.
b. For $5 \leq \lambda_c < 10$, the minimum required plastic moment capacity M_p for the members is amplified by the factor δ_p in a rigid–plastic analysis where $\delta_p = \dfrac{0.9}{1 - \dfrac{1}{\lambda_c}}$.

   The aforementioned equation is equivalent to reducing the plastic collapse load factor by the same factor. Hence, in an elastoplastic analysis, the actual failure load $= \dfrac{\lambda_p}{\delta_p}$.
c. For $\lambda_c < 5$, a second-order plastic analysis (advanced analysis) has to be carried out.

It should be noted that in Eurocode 3, an amplification factor $\left(= \dfrac{1}{1 - \dfrac{1}{\lambda_c}}\right)$ is applied to the loading only when $3 \le \lambda_c < 10$. No specific method is specified in the American AISC design code for amplification calculation due to second-order effects. Finally, it should be noted that a more precise calculation of the actual failure load should be based on Equation (7.19) rather than on Equation (7.22).

Problems

7.1. Determine the plastic collapse load factor α and the plastic rotation demands at the plastic hinges at collapse for the structure shown in Figure P7.1. For the beam, $M_p = 420$ kNm, $E = 2 \times 10^8$ kN/m^2, $I = 320 \times 10^{-6}$ m^4, $A = 0.02$ m^2.

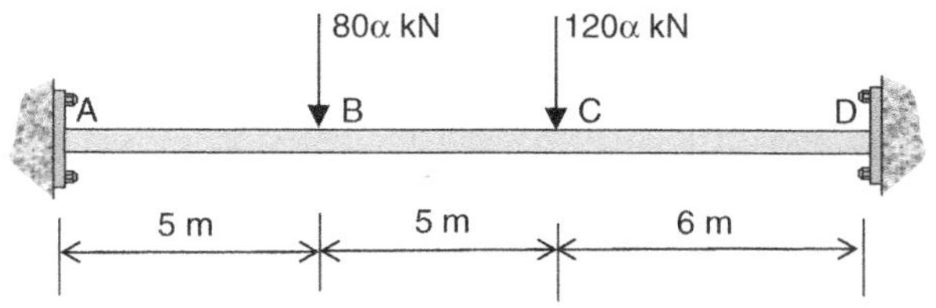

FIGURE P7.1. Problem 7.1.

7.2. Determine the plastic rotation capacity R_m of a 6-m-long simply supported beam of yield stress equal to 250 MPa with a section shown in Figure P7.2. There is no web or warping restraint to the beam. Assume yielding by pure bending only.

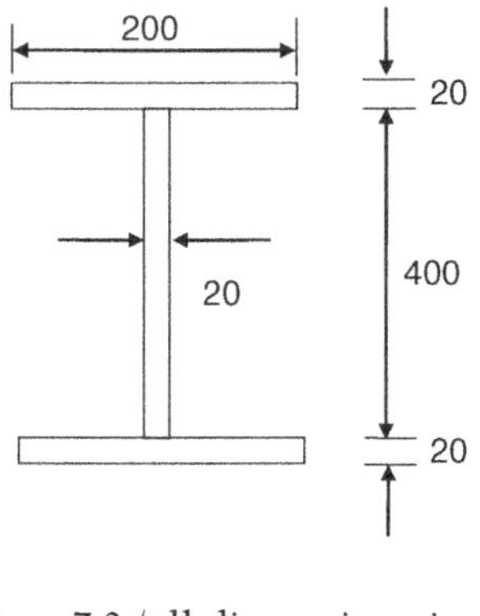

FIGURE P7.2. Problem 7.2 (all dimensions in millimeters).

7.3. Determine the critical temperature of the steel frame shown in Figure P7.3. For the members, $M_p = 558$ kNm, $N_p = 2560$ kN, $\alpha_t = 0.000012/°C$, $E = 2 \times 10^8$ kN/m^2, $I = 477 \times 10^{-6}$ m^4, $A = 0.0105$ m^2. (*Note*: At collapse, the plastic hinges enable the frame to expand freely.)

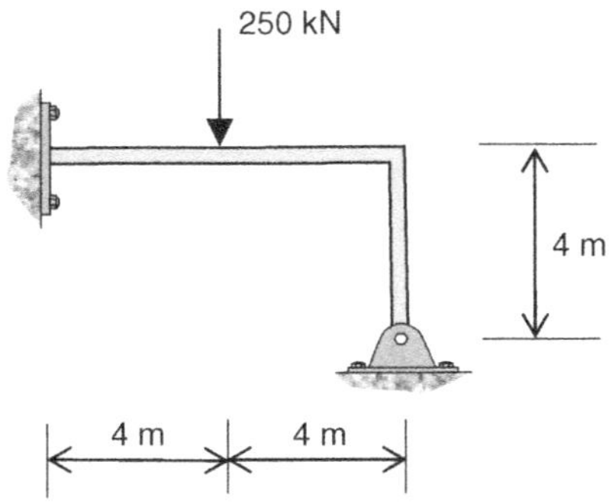

FIGURE P7.3. Problem 7.3.

7.4. Determine the critical temperature of the fixed-end beam ABC made of steel I section shown in Figure P7.4 when (a) the effect of axial force on plastic moment capacity is ignored and (b) the effect of the axial force-bending interaction due to thermal expansion is considered. For the beam, $M_p = 558$ kNm, $N_p = 2560$ kN, $\alpha_t = 0.000012/°C$, $E = 2 \times 10^8$ kN/m^2, $I = 477 \times 10^{-6}$ m^4, $A = 0.0105$ m^2.

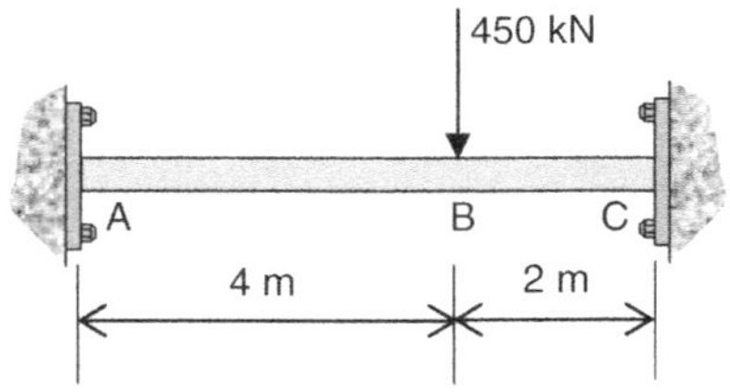

FIGURE P7.4. Problem 7.4.

Bibliography

1. Lay, M. G., and Galambos, T. V. (1967). Inelastic beams under moment gradient. *J. St. Div., ASCE,* **93**(1), pp. 381–399.
2. Kemp, A. R. (1986). Factors affecting the rotation capacity of plastically designed members. *The St. Eng.,* **64B**(2), pp. 28–35.

3. Ziemian, R. D., McGuire, W., and Deierlein, G. G. (1992). Inelastic limit states design, Part I: Planar frame studies. *J. of St. Eng., ASCE,* **118**(9), pp. 2532–2549.
4. Skempton, A. W., and MacDonald, D. H. (1956). Allowable settlement of buildings. *Proc. Inst. Civ. Eng.,* **5**(III), pp. 727–768.
5. Najjar, S. R., and Burgess, I. W. (1996). A nonlinear analysis for three-dimensional steel frames in fire conditions. *Eng. St.,* **18**(1), pp. 77–89.
6. Schleich, J. B., Dotreppe, J. C., and Franssen, J. M. (1985). Numerical simulations of fire resistance tests on steel and composite structural elements or frames. *Proceedings of the First International Symposium, Fire Safety Science,* Oct., pp. 311–323.
7. Wong, M. B. (2001). Elastic and plastic methods for numerical modelling of steel structures subject to fire. *J. Const. St. Res.,* **57**, pp. 1–14.

CHAPTER 8

Design Consideration

8.1 Introduction

In general, most design rules applied to elastic design can be similarly applied to plastic design. Additional rules stipulated in design codes for plastic design are mainly to ensure that plastic moments at yielded sections can be maintained and the plastic collapse load achieved without exhibiting local buckling in the cross sections. Local buckling comes about because of the occurrence of certain structural phenomena when the structure is loaded beyond its linear elastic limit. These structural phenomena, including lateral–torsional buckling, flexural buckling, excessive shear force, and cyclic loading, should be avoided when the plastic design method is used.

One way to ensure that structures perform satisfactorily according to plastic design theory is through laboratory tests. Full-scale tests on the plastic behavior of steel structures have been carried out. Figure 8.1 shows a full-scale test being carried out on a cold-formed steel portal frame at Sydney University.

In the following chapter, major design rules governing the plastic design process are described with reference to Australian (AS4100)[1], European (EC3),[2] and United States (AISC)[3] design codes. For the AISC code, only the load and resistance factor design method is referred to. Both AS4100 and AISC specify that the yield strength of steel for plastic design cannot exceed 450 MPa as ductility becomes a concern when steels of higher yield strength are used.

8.2 Serviceability Limit State Requirements

This is mainly concerned with the deflection of structures. For certain types of structures, such as portal frames subjected to high wind loads,

FIGURE 8.1. Test on cold-formed steel portal frame (courtesy of School of Civil Engineering, the University of Sydney).

horizontal deflection is often excessive and could be the main factor governing the design. The check for serviceability is usually carried out in the same way as for elastic design. That is,

$$\delta \leq \Delta \tag{8.1}$$

where δ is the maximum deflection obtained from elastic analysis under service loads and Δ is the deflection limit.

When the deflection check is performed, a common assumption for plastic design is that no plastic hinge should occur when the structure is subjected to service loads. This assumption is made to avoid the formation of permanent plastic deformation under cyclic loading such as wind. The formation of plastic deformation in a structure under cyclic loading may lead to failure by incremental collapse or alternating plasticity. Incremental collapse occurs when both residual moments and plastic rotations increase after each cycle of loading until a plastic collapse mechanism is formed. Alternating plasticity, also termed low cycle fatigue, is a phenomenon where cracks in the material grow because of repeated plastic deformation until rupture

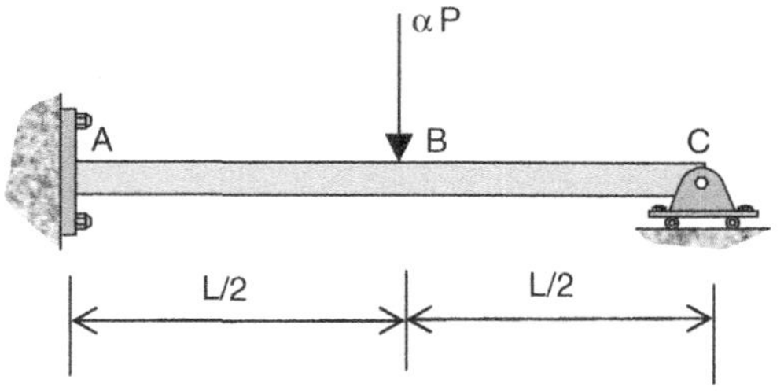

FIGURE 8.2. Beam subjected to cyclic loading.

occurs in the structure. By performing elastoplastic analysis in a step-by-step manner, the behavior of a structure that may fail by incremental collapse or alternating plasticity can be traced.

Take the propped cantilever beam shown in Figure 8.2 as an example. For this structure, the collapse load factor α_c is 1.35. Suppose that a cyclic load with a maximum value of $\alpha = 1.3$ for $P = 10$ kN is applied to the structure. $L = 12$ m and $M_p = 27$ kNm.

The application of the load follows the following sequence in the elastoplastic analysis: 1.3P, −1.3P, −1.3P, 1.3P. The sequence represents a complete cycle of the load applied in both upward and downward directions. In the first cycle, only one plastic hinge occurs at A. The plastic rotation of the plastic hinge at A, which is unloaded after the peak load is reached, is also calculated. Results of the first half-cycle of loading are shown in Table 8.1. At the end of the first half-cycle of loading (after stage 3 analysis), the applied load is reduced to zero, leaving a set of self-equilibrating residual moments in the beam and permanent deformation of $1080/EI$ at A.

TABLE 8.1
Results of the first half-cycle of loading

Analysis Stage No: 1 — Critical Load Factor, $\alpha_{cr} = 1.2$

Member	*Joint*	*Moment* M_o	*Residual plastic moment* $M_p - M_i$	*Load factor* $\alpha = \frac{M_p - M_i}{M_o}$	*Cumulative moment* $M_{i+1} = M_i + \alpha_{cr} M_o$	*Permanent rotation at A (radian)*
AB	A	22.5	27	1.2	27	
	B	18.75	27	1.44	22.5	
BC	B	−18.75	−27	1.44	−22.5	
	C	0	27	1.44	0	

Analysis Stage No: 2 Load Factor, $\alpha = 0.1$

Member	*Joint*	*Moment* M_o	*Residual plastic moment* $M_p - M_i$	*Load factor* $\alpha = \frac{M_p - M_i}{M_o}$	*Cumulative moment* $M_{i+1} = M_i + \alpha_{cr} M_o$	*Permanent rotation at A (radian)*
AB	A	0	0		27	1080/*EI*
	B	30	4.5		25.5	
BC	B	−30	−4.5		−25.5	
	C	0	27		0	

Analysis Stage No: 3 Load Factor, $\alpha = -1.3$

Member	*Joint*	*Moment* M_o	*Residual plastic moment* $M_p - M_i$	*Load factor* $\alpha = \frac{M_p - M_i}{M_o}$	*Cumulative moment* $M_{i+1} = M_i + \alpha_{cr} M_o$	*Permanent rotation at A (radian)*
AB	A	−22.5	−54		−2.25	1080/*EI*
	B	−18.75	−52.5		1.125	
BC	B	18.75	52.5		−1.125	
	C	0	27		0	

Analysis Stage No: 4 Load Factor, $\alpha = -1.1$

Member	*Joint*	*Moment* M_o	*Residual plastic moment* $M_p - M_i$	*Load factor* $\alpha = \frac{M_p - M_i}{M_o}$	*Cumulative moment* $M_{i+1} = M_i + \alpha_{cr} M_o$	*Permanent rotation at A (radian)*
AB	A	−22.5	−24.75	1.1	−27	1080/*EI*
	B	−18.75	−28.125	1.5	−19.5	
BC	B	18.75	28.125	1.5	19.5	
	C	0	27		0	

Analysis Stage No: 5 Load Factor, $\alpha = 0.2$

Member	*Joint*	*Moment* M_o	*Residual plastic moment* $M_p - M_i$	*Load factor* $\alpha = \frac{M_p - M_i}{M_o}$	*Cumulative moment* $M_{i+1} = M_i + \alpha_{cr} M_o$	*Permanent rotation at A (radian)*
AB	A	0	54		−27	−1080/*EI*
	B	−30	−7.5		−25.5	
BC	B	30	7.5		25.5	
	C	0	27		0	

Analysis Stage No: 6 Load Factor, $\alpha = 1.3$

Member	Joint	Moment M_o	Residual plastic moment $M_p - M_i$	Load factor $\alpha = \frac{M_p - M_i}{M_o}$	Cumulative moment $M_{i+1} = M_i + \alpha_{cr} M_o$	Permanent rotation at A (radian)
AB	A	22.5	54		2.25	1080/*EI*
	B	18.75	52.5		−1.125	
BC	B	−18.75	−52.5		1.125	
	C	0	27		0	

At the end of stage 6 calculations, the load is again reduced to zero, leaving a set of residual, self-equilibrating moments as shown. By repeating the aforementioned process, it can be shown that the residual moments do not grow. This structure exhibits alternating plasticity under the cyclic load. A cycle of the load–plastic rotation at A is shown in Figure 8.3.

8.3 Ultimate Limit State Requirements

Ultimate limit state requirements for plastic design stipulated in most design codes are based on the assumption that structures using this design method are ductile enough to sustain plastic deformation without premature failure. Most rules for plastic design were developed

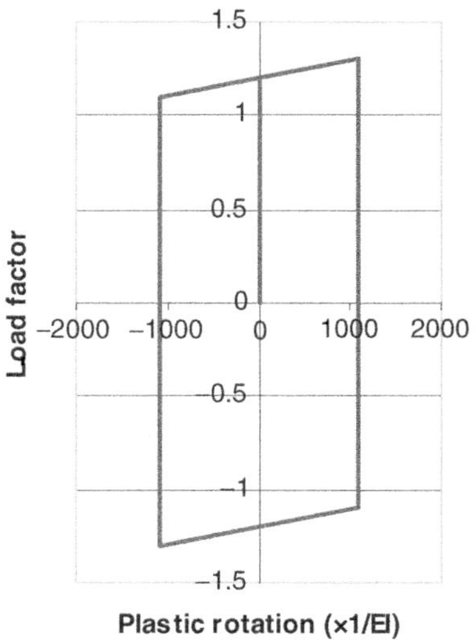

FIGURE 8.3. Load–plastic rotation relationship for one cycle of load at A.

many years ago when rigid–plastic theory was commonly used for analyzing and designing simple to moderately complex structures. The elastoplastic analysis method introduced in this book enables the design of structures with virtually any degree of complexity. Therefore, care must be taken when applying these rules to complex structures. For some complex structures, it may be necessary to check rotational capacity in plastic hinges as described in Chapter 7 even after all design code requirements are satisfied. Because of the complicated interaction between yielding of steel material and local buckling, most design rules, many of them empirical, apply specifically to standard structural sections with double or monosymmetry such as I sections, box sections, channels, and circular hollow sections.

8.3.1 Class of Sections

The ability of a plate element with specified end support conditions to yield under compression before buckling occurs depends on the yielding strength and the slenderness of the cross section of the plate. The slenderness is defined as the ratio of its length (b) to its thickness (t) in the cross section of the plate element shown in Figure 8.4.

To ensure that a section is able to sustain plastic moment without exhibiting local buckling in plastic design, the slenderness ratios of all plate elements in the cross section must be less than some limiting values. In AS4100 and AISC, structural steel cross sections are classified into three categories—compact, noncompact, and slender—whereas in EC3, structural steel cross sections are classified into four categories—classes 1, 2, 3, and 4. For plastic design, AS4100 and AISC require that sections of the structural members must be compact, whereas EC3 requires the use of class 1 or class 2 sections. Class 2 sections are used at the location only where the last plastic hinge occurs. For other section classes in EC3, rotation capacity should be checked against plastic rotation demand for plastic hinges. In all three design codes, the slenderness must be less than a value related to μ as explained later. The ways slenderness is expressed to satisfy plastic design in the three design codes are given in Table 8.2.

Slight differences in defining slenderness of plate elements in the cross sections exist among the three design codes. The definitions of the dimension b for calculating the slenderness of the plate elements of both welded and rolled I sections are shown in Figure 8.5.

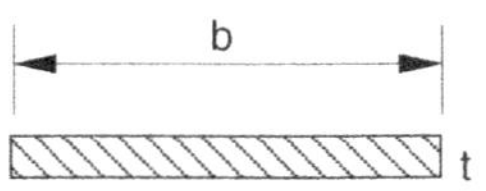

FIGURE 8.4. Slenderness ratio of a plate $= b/t$.

TABLE 8.2
Slenderness requirements for plastic design

	AS4100	EC3	AISC
Slenderness limit	$\frac{b}{t}\sqrt{\frac{f_y}{250}} \le \mu$	$\frac{b}{t} \le \mu\sqrt{\frac{235}{f_y}}$	$\frac{b}{t} \le \mu\sqrt{\frac{E}{f_y}}$

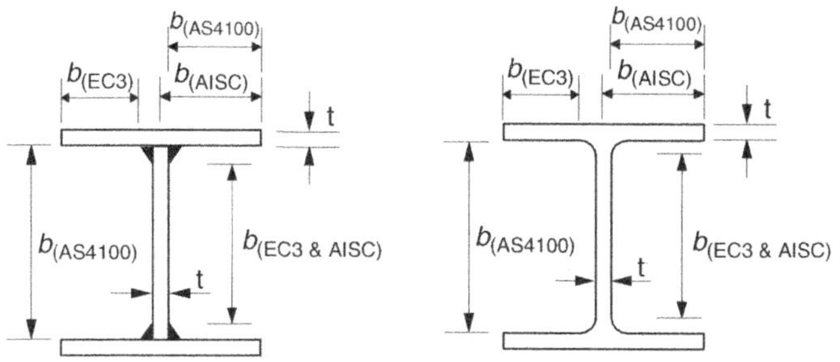

FIGURE 8.5. Values of b in different codes.

For plates under compression as a consequence of bending, the values of μ for different loading conditions are given in Table 8.3.

For comparison, slendernesses for a section with $f_y = 300$ MPa and $E = 200000$ MPa are calculated using the rules in the three codes and the results are shown in Table 8.4.

TABLE 8.3
Values of μ

	AS4100	EC3	AISC
Flange (compression by bending)	9	9	0.38
Web (bending)	82	72	3.76
Web (compression only)	30	33	1.12

TABLE 8.4
Limiting slenderness ratio (b/t) for plate elements in I-sections

	AS4100	EC3	AISC
Flange (compression by bending)	8.2	8.0	9.8
Web (bending)	74.9	63.7	97.1
Web (compression only)	27.4	29.2	28.9

8.3.2 Unbraced Length

To prevent lateral–torsional buckling from occurring in a beam that contains plastic hinges, a maximum unbraced length between lateral braces along the length of the beam subject to bending should be specified. The member should be restrained by lateral braces along the compression flange (torsional restraints). The following requirements apply only to double symmetric I sections.

AS4100

The unbraced length L_b between lateral braces cannot exceed a length given by

$$\frac{L_b}{r_y} \leq (80 + 50\beta_m)\sqrt{\frac{250}{f_y}} \tag{8.2}$$

where r_y is the radius of gyration about the minor axis and β_m is the ratio of smaller to larger end moments in the length L_b, positive for reverse curvature and negative for single curvature.

EC3

The maximum unbraced length between torsional restraints of a member segment subject to bending is determined by satisfying the following empirical formula:

$$\frac{L_b}{r_y} \leq \frac{\left(5.4 + \frac{600 f_y}{E}\right)\left(\frac{h}{t_f}\right)}{\sqrt{5.4\left(\frac{f_y}{E}\right)\left(\frac{h}{t_f}\right)^2 - 1}} \tag{8.3}$$

where h is the depth of the section and t_f is the flange thickness. Equation (8.3) is valid provided that there are one or more intermediate restraints between the torsional restraints at a spacing satisfying the requirement for L_m given by Equation (8.12) in Section 8.3.4.2 with $N^* = 0$.

AISC

Both AS4100 and AISC have similar conditions for calculating the maximum unbraced length. The condition for AISC is given by

$$\frac{L_b}{r_y} \leq [0.12 + 0.076\beta_m]\left(\frac{E}{f_y}\right) \tag{8.4a}$$

Equation (8.4a) is similar to that in an earlier version of AISC rules in which the modulus of elasticity E in Equation (8.4a) is multiplied into the equation. If $E = 30000$ ksi, Equation (8.4a) then becomes

$$\frac{L_b}{r_y} \leq [3600 + 2280\beta_m]\left(\frac{1}{f_y}\right) \tag{8.4b}$$

8.3.3 Plastic Hinge Stability

The region close to a plastic hinge in a member is vulnerable to lateral buckling instability because of the rotational nature of the plastic hinge. A common means to ensure stability in a region where a plastic hinge occurs is to provide adequate lateral restraints in that region. Both flanges of the section should be restrained.

AS4100

Lateral restraints are provided within a distance of $0.5d_1$ from the plastic hinge and d_1 is the clear depth of the web. In addition, if a design bearing load, P_b^* (such as a secondary beam sitting on a main beam), or a design shear force, V^*, is such that

$$P_b^* \text{ or } V^* \geq \frac{V_c}{10} \tag{8.5}$$

where V_c is the shear capacity of the web, then load bearing stiffeners should be provided as shown in Figure 8.6.

EC3

Lateral restraints should be provided within a distance of 1.5 times the flange width, or 0.5 times the height of the plastic section, from the plastic hinge.

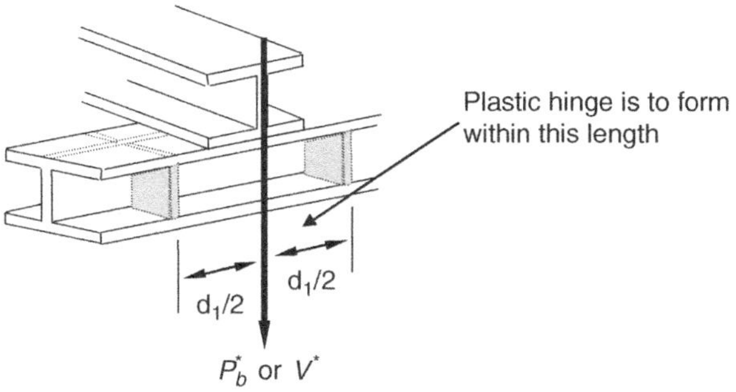

FIGURE 8.6. Shear requirements at a plastic hinge.

AISC

No additional recommendation is made other than that the member should satisfy the maximum unbraced lateral restraint length requirement.

8.3.4 Combined Actions Requirements

In plastic design, the effect of the combined actions of bending moment and axial force is most significant. The following requirements ensure that the plate elements in the cross section do not cripple due to axial force while the section is undergoing plastic bending. The requirements apply mainly to steel members made of I sections.

AS4100

Member Slenderness

To ensure that the full moment capacity can be maintained when the collapse mechanism develops, check the axial force N^* where

$$\frac{N^*}{\phi N_s} \leq \left[\frac{0.60 + 0.40\beta_m}{\sqrt{N_s/N_{OL}}}\right]^2 \quad \text{when } \frac{N^*}{\phi N_s} \leq 0.15 \tag{8.6a}$$

and

$$\frac{N^*}{\phi N_s} \leq \left[\frac{1 + \beta_m - \sqrt{N_s/N_{OL}}}{1 + \beta_m + \sqrt{N_s/N_{OL}}}\right] \quad \text{when } \frac{N^*}{\phi N_s} > 0.15 \tag{8.6b}$$

where ϕN_s is the design squash load of the section with capacity factor $\phi = 0.9$,

$$N_{OL} = \frac{\pi^2 EI}{L^2}$$

where L is the actual length of the member.

If Equation (8.6b) is not satisfied, the member shall not contain plastic hinges and should be designed elastically.

Web Slenderness

To ensure that no local buckling occurs to the web due to both bending moment and axial force interaction, check the axial force N^* along the length of the member where

$$\frac{N^*}{\phi N_s} \leq 0.60 - \left[\frac{d_1}{t_w}\frac{\sqrt{(f_y/250)}}{137}\right] \quad \text{for webs where } 45 \leq \frac{d_1}{t_w}\sqrt{\frac{f_y}{250}} \leq 82 \tag{8.7a}$$

or

$$\frac{N^*}{\phi N_s} \leq 1.91 - \left[\frac{d_1}{t_w}\frac{\sqrt{(f_y/250)}}{27.4}\right] \leq 1.0 \quad \text{for webs where } 25 < \frac{d_1}{t_w}\sqrt{\frac{f_y}{250}} < 45 \tag{8.7b}$$

or

$$\frac{N^*}{\phi N_s} \leq 1.0 \qquad \text{for webs where } \frac{d_1}{t_w}\sqrt{\frac{f_y}{250}} \leq 25 \tag{8.7c}$$

where d_1 is the clear web height as defined in Figure 8.5 and t_w is the web thickness.

For the case where $\frac{d_1}{t_w}\sqrt{\frac{f_y}{250}} > 82$ (local buckling occurs under pure bending), the member must be designed elastically.

EC3

Details of the relevant clauses in EC3 in relation to combined actions for members containing plastic hinges have been explained by King.[4] The theory was published by Horne[5] on elastic stability of columns under the actions of axial force N^* and bending moment M^*. It is based on the following simple combined action equation:

$$\left(\frac{M^*}{M_{cr}}\right)^2 + \frac{N^*}{N_{cr}} = 1 \tag{8.8}$$

where M_{cr} is the critical bucking moment of the member with no warping stiffness given by

$$M_{cr} = \sqrt{\left(\frac{\pi^2 EI}{L^2}\right) GI_t} \tag{8.9}$$

$$N_{cr} = \frac{\pi^2 EI}{L^2} \tag{8.10}$$

where G is shear modulus, L is length of member, and I_t is torsion constant for the cross section $= \sum(bt^3/3)$.

For stable sections with plastic hinges within the laterally restrained length L_m, M_{cr} is equal to the elastic, or conservatively the plastic, moment capacity moderated by a moment gradient factor C_1 provided that L_m is not greater than $0.6L$. That is,

$$M^* = M_{cr} = \frac{W_{pl} f_y}{C_1} \tag{8.11}$$

where W_{pl} is the plastic section modulus of the section. The values of C_1 for different cases of moment gradient can be found in EC3. Using Equations (8.8) to (8.11), it can be found that the limiting value of L_m is

$$\frac{L_m}{r_y} = \frac{38}{\sqrt{\frac{1}{57.4}\left(\frac{N^*}{A}\right) + \frac{756}{C_1{}^2}\left(\frac{W_{pl}^2}{AI_t}\right)\left(\frac{f_y}{235}\right)^2}} \tag{8.12}$$

AISC

For web slenderness, check

(i) For $\dfrac{N^*}{\phi N_s} \leq 0.125$,

$$h/t_w \leq 3.76\sqrt{\frac{E}{F_y}}\left(1 - 2.75\frac{N^*}{\phi N_s}\right) \tag{8.13a}$$

(ii) For $\dfrac{N^*}{\phi N_s} > 0.125$,

$$h/t_w \leq 1.12\sqrt{\frac{E}{F_y}}\left(2.33 - \frac{N^*}{\phi N_s}\right) \geq 1.49\sqrt{\frac{E}{F_y}} \tag{8.13b}$$

where h is the clear distance between the flanges. For a column designed by the plastic method,

$$\phi N_s \leq 0.85(\phi f_y A) \quad \text{for braced frames} \tag{8.14a}$$

$$\phi N_s \leq 0.75(\phi f_y A) \quad \text{for moment frames} \tag{8.14b}$$

In addition, the slenderness of the column has to satisfy the following:

$$L/r \leq 4.71\sqrt{\frac{E}{F_y}} \tag{8.15}$$

8.3.5 Connections

In general, a full-strength connection should have a strength capacity not less than that of the connecting members. This assumption has been made when connections are designed in accordance with elastic design.

The rotation capacity at any of the plastic hinges in the connections should not be exceeded. If the materials of the steel members satisfy the ductility requirements, it is unlikely that the rotation capacity of any plastic hinge in a simple or moderately complex structure will be exceeded when all other design requirements are satisfied.

Bibliography

1. Standards Australia (1998). *AS4100-1998, Steel structures.* SA, Australia.
2. EC3 (2002). *prEN 1993-3. Eurocode 3: Design of steel structures, Part 3: Buildings.* CEN.
3. ANSI/AISC 360-05 (2005). *Specification for structural steel buildings.* American Institute of Steel Construction.
4. King, C. (2005). Member stability at plastic hinges: The background to Annex BB.3. *Proceedings, Eurosteel 2005, 4th European Conference on Steel and Composite Structures.* Hoffmeister, B., and Hechler, O., **Volume B, 3-25 to 3-32**.
5. Horne, M. R. (1964). Safe loads on I-section columns in structures designed by plastic theory. *Proc. Inst. Civ. Eng.,* **29**.

Answers

Chapter 1

1.1. 1
1.2. 1
1.3. 1
1.4. 16
1.5. $\theta_A = 0.0357$ rad., $\theta_B = -0.0714$ rad.
1.6. Axial: AB = −418.7 kN, BC = −295.3 kN

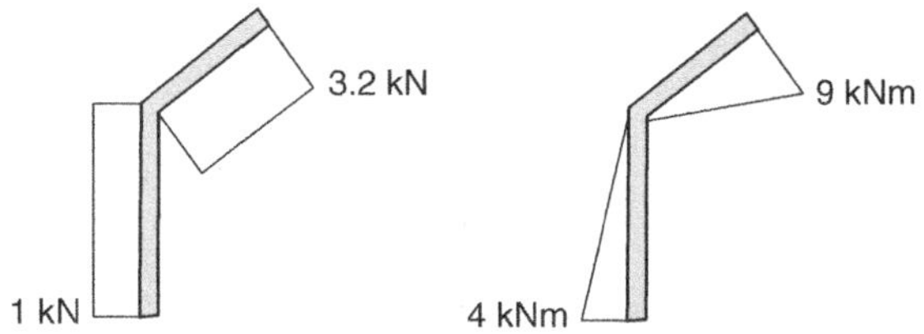

SF diagram BM diagram

1.7.

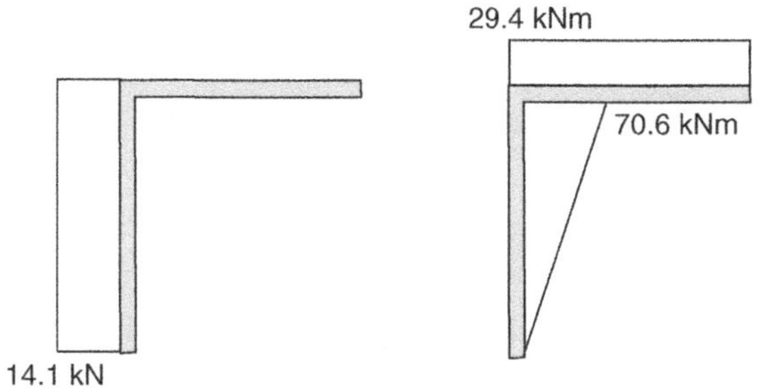

SF diagram BM diagram

1.8. $\theta_A = -0.111$ rad., $\theta_B = -0.0864$ rad. $v_B = -0.181$ m.

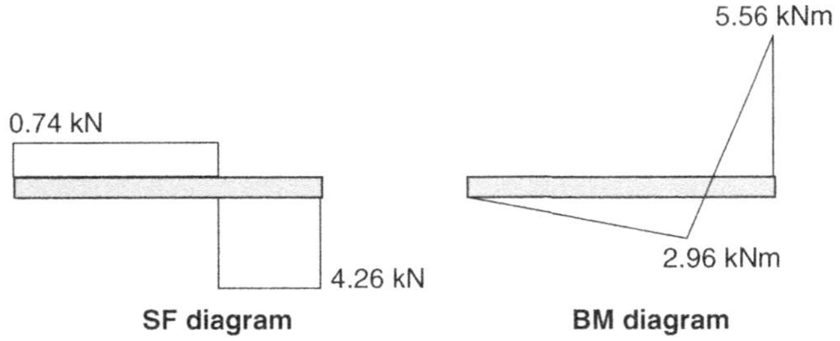

1.9. Axial force: AB = 43.1 kN, BC = 48.1 kN
Shear force: AB = 28.4 kN, BC = 2.85 kN
Bending moment: AB = 26.7 kNm, BA = BC = 9.7 kNm, CB = 4.5 kNm

1.10. n k

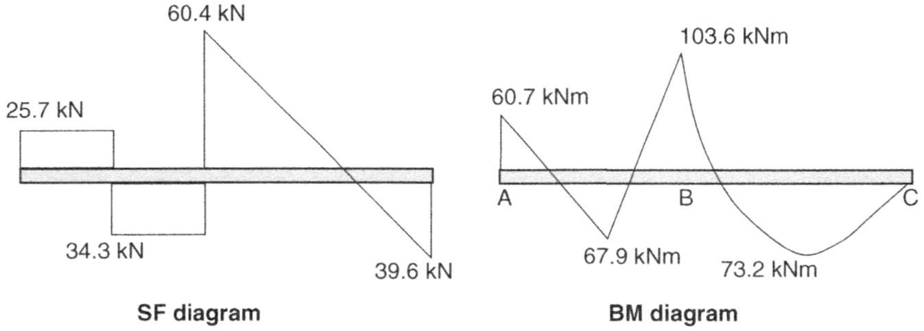

1.11. EA = 1.170×10^9 N, EI = 2.653×10^{13} Nmm^2.

Chapter 2

P2.2. $x_p = 4.619$ m

P2.3. (a) $x_{p1} = 0.968$ m, $x_{p2} = 1.455$ m
(b) $x_{p1} = 0.833$ m, $x_{p2} = 2.5$ m

P2.4. (a) $x_{p1} = 0.285$ m
(b) $x_{p1} = 0.25$ m; $x_{p2} = 0.667$ m; $x_{p3} = 0.417$ m

P2.5. (a) y = 600 mm; Plastic moment = 1249 kNm
(b) y = 700 mm; Plastic moment = 1045 kNm

P2.6. 276 mm from bottom; $Z_s = 3.05 \times 10^6$ mm^3

Chapter 3

P3.1. (a) P = 40 kN; (b) P = 30 kN
P3.2. (a) $\alpha = 8$; (b) $\alpha = 6.78$

Chapter 4

P4.1. $\alpha = 2.0$ (first hinge at $\alpha = 1.896$)
P4.2. $\alpha = 1.152$; deflection at B = 98.2 mm; at design load level, moment = −90.7 kNm, deflection = 62.6 mm
P4.3. a) Answers will be used for part c;
b) $\alpha_{col} = 1.371$; c) $M^* = 126.2$ kNm, $N^* = 331.6$ kN;
d) $M^* < 293$ kNm
P4.4. $\alpha = 196.2$ (exact)
P4.5. $\alpha = 0.8293$ after 4 iterations
P4.6. P = 112.3
P4.7. P = 889.4 with deflection = 318 mm at C
P4.8. $\alpha = 1.259$ with hinges at A and at 6.423 m from A
P4.9. A, D, G, H. $\alpha = 1.760$

Chapter 5

P5.1. $\alpha = 2.0$
P5.2. $M_p = 70$ kNm
P5.3. $\lambda = 1.47$
P5.4. $\alpha = 1.26$
P5.5. $M_p = 80$ kNm
P5.6. $P = \frac{7M_P}{6L}$
P5.7. $\alpha = 0.829$. Plastic hinges formed at mid-span and right end of the rafter.
P5.8. $P = \frac{13M_P}{6L}$ with plastic hinges at A, B and D.

Chapter 7

P7.1. $\alpha_c = 1.400$, hinges at D, A, C; at $\alpha = 1.164$,
$\theta_D = 4.53 \times 10^{-4}$ radians; at $\alpha = 1.400$, $\theta_D = 1.12 \times 10^{-2}$ radians, $\theta_A = 1.066 \times 10^{-2}$ radians.
P7.2. $l_f = 0.182$, $R_m = 4.02$
P7.3. 596 °C
P7.4. (a) $T_c = 534$ °C; (b) $T_c = 76$ °C

Index

A

accidental loadings, design codes, 75
AISC, 76, 215, 219, 224–31
  see also United States
allowable stress design (ASD), concepts, 74–6
AS4100, 76, 77–8, 219, 224–31
  see also Australia
ASD *see* allowable stress design
Australia
  see also AS4100
  design codes, 75–6, 86–7, 207–8, 214, 219, 224–31
automated analysis, spreadsheets, 111, 112–16, 175–92, 209
axial forces
  see also design actions; yield surface
  combined actions requirements, 228–30
  concepts, 22–32, 42–53, 60–9, 82–7, 91–2, 94–5, 99–106, 116–24, 133–7, 157–9, 175–93, 206–12, 228–30
  diagram sign convention, 22–7
  moment–curvature relationship, 60–9, 227–30
  plastic collapse load effects, 157–9
  reduced plastic moment capacity, 83–7, 119–24, 157–9
  temperature effects, 45–50, 206–12
axial–bending interaction
  concepts, 87, 94–5, 99–106, 133–7, 159, 175–93, 197–200, 205–6, 228–30
  pure bending, 99–106, 119–24, 175–93, 197–200, 205–6, 228–30

B

basic mechanisms
  see also complete collapse; partial collapse
  concepts, 151–5, 164
beam mechanisms, partial/complete collapse, 152–9
beams, 1, 34–45, 50–3, 56–80, 92–5, 109–12, 130–1, 139–62, 167–75, 184–9, 202–3, 208–12
bending moments
  see also design actions; yield surface
  concepts, 22–32, 48–50, 51–3, 55, 57–69, 72–4, 108–24, 130–7, 140–4, 157–9, 166–75, 213–16, 228–30
  diagram convention, 22–7, 140, 142–4
  elastic–plastic behavior, 57–69, 72–4, 92–5, 196–8
  load factor multiplications, 109–11
  moment–curvature relationship, 59–72, 227–30
  second-order effects, 76, 213–16
bending–axial interaction
  see axial–bending...
bending–shear interactions, 86–7

C

cantilever beams, 34–5, 37–8, 67–9, 109–12, 130–1, 143, 149–51, 159–62, 167–75, 184–9, 202–3, 208–12, 221–3
class of sections, 224–6
coefficient of linear expansion, temperature effects, 46–50, 208–12
cold-formed steel portal frames, 219–20
collapse, 2, 58–9, 72–3, 76, 81–3, 99–106, 107–37, 139–62, 163–92, 195–217, 220–30
  affecting factors, 195–217
  complete collapse, 147–55
  foundation settlements, 195, 200–17
  partial collapse, 147–51
  problems, 215–16
  second-order effects, 76, 213–16
  temperature effects, 45–50, 53, 75, 195, 206–12

combined actions requirements, design issues, 228–30
combined beam/sway mechanisms, partial/complete collapse, 152–9, 164, 210–11
common loading cases, 28–32, 34–45, 75, 107–37
compact sections, plastic designs, 76–7
compatibility condition, structure stiffness matrix, 14–16
complete collapse
  see also collapse
  continuous beams and frames, 147–55
  definition, 148
compression
  axial forces, 22–7, 45–50, 61–2, 69–72, 83–7
  buckling, 224–30
  temperature effects, 45–50, 206–12
computational issues
  moment–curvature relationship, 65–9
  structural analysis, 1–2
computer programs, 1–3, 66–9, 76, 107, 108–16, 119–20, 122–3, 164, 166–92, 209
  see also spreadsheet applications
  concepts, 1–3, 76, 108–16, 119–20, 122–3, 164, 166–75, 183–9
  elastoplastic analysis, 108–16, 119–20
  incremental elastoplastic analysis, 108–16, 119–20
  MATLAB, 164, 175, 183–9
  plastic analysis, 1, 76, 108–16, 164, 166–75
concrete, ductility, 196
condensation method
  see also member stiffness matrix; pins
  cases, 33–40
  concepts, 32–45, 51, 198–200
  plastic rotation demand, 198–200
  procedures, 41
connections, concepts, 230
constrained optimisation problems, linear programming, 164–93
continuous beams
  concepts, 50–1, 76, 147–55
  partial collapse, 147–51
  UDL on end span, 149–51, 160
coordinates
  global coordinate system, 3, 8–9, 11–14, 28–32, 35–45
  local coordinate system, 8–9, 11–14
  transformation concepts, 11–13, 28–32
critical temperature
  see also temperature...
  concepts, 207–12, 216
  determination, 208–9, 216
  elastoplastic analysis, 211–12
crookedness, second-order effects, 76, 213–16
cross-sectional area
  elastic–plastic behavior, 55–69, 195
  plastic designs, 76–9
  reduced plastic moment capacity due to force interaction, 83–7, 119–24, 157–9
  temperature effects, 46–50, 206–12
  yield surface, 87–92
cyclic loading, 219–21

D

deflection, 10–11, 58–9, 72–3, 75, 100–6, 107–37, 219–30
  calculations, 116–18
  serviceability limit state, 75, 219–30
deformation, 58–9, 92, 100–6, 119–24, 175–93, 220–30
degrees of freedom, concepts, 3–9, 28–32, 192–3
design actions
  see also axial forces; bending moments; shear forces
  calculations, 116–18
  concepts, 116–18, 155–7, 213–16
  elastoplastic analysis, 116–18
  linear interpolation, 117–18, 155, 212
  types, 116
design codes, 46, 75–6, 86–7, 197–8, 207–16, 219–31
  accidental loadings, 75
  Australia, 75–6, 86–7, 207–8, 214, 219, 224–31
  concepts, 75–6, 86–7, 207–16, 219–31
  ductility needs, 75–6, 219, 223–30
  Europe, 75–6, 219, 224–31
  local buckling, 197–8, 213–14, 219–31
  overview, 75–6
  plastic designs, 75–6
  second-order effects, 76, 213–16
  United States, 74–6, 215, 219, 224–31
design issues
  see also plastic designs
  ASD, 74–6
  class of sections, 224–6
  combined actions requirements, 228–30

concepts, 75–6, 86–7, 116–18, 155–7, 207–16, 219–31
connections, 230
cyclic loading, 219–21
full-scale tests, 219–20
LSD, 74–5, 219–30
member slenderness, 228–30
optimal plastic design problem, 164, 189–93
plastic hinge stability, 227–8
serviceability limit state requirements, 219–30
slenderness, 224–30
strain hardening, 57, 73–4, 197–200
unbraced length, 226–7
web slenderness, 228–30
deterioration of steel, temperature effects, 45–50, 53, 206–12
direct method, force interaction effects on plastic collapse, 119–22, 203–6
direct stiffness method
see also general…; stiffness…
concepts, 3, 6–9
global coordinate system, 8–9, 11–14
local coordinate system, 8–9, 11–14
statical indeterminacy, 3, 6–9, 20
discrete plane frame problem, 164, 175–93
displacement method *see* stiffness method
displacements
concepts, 2–18, 21–7, 31–2, 35–45, 51–3, 81–3, 146–7, 195–217
second-order effects, 76, 213–16
structural analysis, 2–18, 21–7, 31–2, 35–45, 51–3
transformations, 12–13
dissipative mechanical energy, plastic flow theory, 91–2
distributed loads
concepts, 126–30, 144–7
elastoplastic analysis, 126–30
dual LP statements, discrete limit analysis problem, 182–3
ductility
see also plastic rotation capacity
concrete, 196, 219, 223–30
design codes, 75–6, 219, 223–30
glass, 196
material types, 195–6
plastic designs, 75–6, 77–9, 195–217, 219, 223–30
steel, 196–200, 223–30
dynamic analysis, concepts, 55

E

EC3, 219, 224–31
see also Europe
eccentric loads, second-order effects, 76, 213–16
elastic analysis
concepts, 1–2, 72, 73–4, 81–3, 92–5, 108–16, 197–200
definition, 72
elastic approach
concepts, 1–2, 55–69, 72, 73–4, 81–3, 92–5, 108–16, 197–200
definition, 55
plastic analysis, 1, 55–69
elastic behavior, moment–curvature relationship, 59–62, 227–30
elastic designs, plastic designs, 73–4, 76, 197–200, 219–20
elastic–plastic behavior
concepts, 55–72, 78, 90–106, 196–8, 227–30
moment–curvature relationship, 59–72, 78, 227–30
elastic–plastic–elastic states, 90–1
elastoplastic analysis
see also elastic…; extrafreedom method; plastic…
computer programs, 108–16, 119–20, 122–3
concepts, 2–3, 32, 55, 58–72, 81–106, 107–37, 203–6, 211–12, 224–30
critical temperature evaluation, 211–12
definition, 55, 72
design actions, 116–18
design issues, 116–18, 219–30
direct method, 119–22, 203–6
distributed loads, 126–30
hyperplanes, 105, 119–24
incremental aspects, 68–9, 81–3, 91–2, 107–37
moment–curvature relationship, 59–72, 78, 227–30
pins, 32, 82–3, 99–102
plastic flow rule, 81–106
plastic hinge concept, 68–9, 72–3, 76–9, 81–106, 107–37
propped cantilever beams, 109–12, 130–1, 208–12, 221–3
spreadsheet applications, 108–16
stiffness matrix, 99–101
yield surface, 92–9, 104–5, 119–24
elastoplastic stiffness matrix
concepts, 91–106, 119–24

elastoplastic stiffness matrix *(continued)*
derivation, 92–5
end moment signs, 97–9, 103–4
forms, 98–9
sections, 95–9
equilibrium condition, structure stiffness matrix, 14, 16–18, 20–7
Eurocode, 3, 76
Euromech Colloquium at Imperial College, 164
Europe
see also EC3
design codes, 75–6, 219, 224–31
expansion, temperature effects, 45–50, 206–12, 216
external loads
internal loads, 27–32, 126–30
joints, 2–27, 112–16
structural analysis, 1–54
external work, UDL calculation method, 146–7
extrafreedom method
see also elastoplastic analysis; pins
concepts, 32, 41–5, 51, 198–200

F

Faculté Polytechnique de Mons, 164
failure temperatures, 195, 206–12
see also temperature...
finite element method, 207
see also direct stiffness...
fixed joints, statical indeterminacy, 4–7, 30–2
fixed-end forces
see also modified end actions
common loading cases, 28–32, 34–45, 107–37
concepts, 27–32, 34–50, 101–2, 142–7, 160–2, 199–200, 203–6, 216
internal loads, 27–32, 34–50, 101–2
flexibility method, structural analysis, 2, 142–4
flexural buckling, 219–31
force interaction
concepts, 2, 83–7, 119–24, 157–9
direct method, 119–22, 203–6
indirect method, 119, 211–12
plastic collapse effects, 119–24
reduced plastic moment capacity, 83–7, 119–24, 157–9
successive approximation method, 119, 122–4, 157–9
force method *see* flexibility method
force resolution, 11–12
Foulkes' classical theorems, 189
foundation settlements
concepts, 195, 200–17
modified end actions, 203–6
plastic analysis, 201–3
fractures, elastic–plastic behavior, 56–69
frames, 1–6, 28–32, 51–2, 55–6, 76, 90–2, 113–24, 128–30, 139–62, 164–93, 203–6, 209–12, 219–20, 230
see also pins
partial collapse, 147, 151–5
statical indeterminacy, 4–6, 30–2
free-body diagrams, 155–7
freedom codes
concepts, 3–6, 8–13, 18–45
extrafreedom method, 32, 41–5, 198–200
pins, 32–45
full-scale tests, 219–20
fully plastic section, moment–curvature relationship, 69–73, 78–9, 107–37

G

gable frames, 134–5, 172–5
GAMS, 183
Gaussian elimination method, 19–20
general elastoplastic stiffness matrix
see elastoplastic stiffness matrix
general stiffness method
see also direct...
concepts, 3
geometric effect, 75
glass, ductility, 196
global coordinate system
concepts, 3, 8–9, 11–14, 28–32, 35–45
direct stiffness method, 8–9, 11–14
member stiffness matrix, 13–14, 15–18, 30–2, 35–45

H

Hooke's Law, 6–7
hyperplanes, 105, 119–24, 179–83

I

increasing loads, elastic–plastic behavior, 55–69, 72–3, 100–6, 107–37, 195
incremental elastoplastic analysis
see also elastoplastic analysis
computer programs, 108–16, 119–20, 122–3
concepts, 68–9, 81–3, 91–2, 107–37, 141

distributed loads, 126–30
linear elastic analysis, 108–16
problems, 130–7
spreadsheet applications, 108–16
indeterminacy concepts, 2–6, 20, 143–4, 148–55
see also complete collapse; partial collapse
degrees, 2, 4–6, 30–2, 50, 143–4, 148, 153–5
indirect method, force interaction effects on plastic collapse, 119, 211–12
inelastic zone, concepts, 68–9, 78–9
internal force, 2, 7–9
internal loads
concepts, 2, 27–32, 33–50, 126–30
fixed-end forces, 27–32, 34–50, 101–2
temperature effects, 27, 45–50, 206–12
treatment, 27–32
types, 27–8
International Centre for Mechanical Sciences, 164, 183
Internet downloads, GAMS, 183
iterative Gauss–Seidel method, 19–20
iterative methods, successive approximation method, 107, 119, 122–4, 157–9

J

joints
concepts, 2–6, 8, 27, 112–16, 126–30, 136–7
external loads, 2–27, 112–16
statical indeterminacy, 4–6, 30–2

K

kinematic theorem (upper bound theorem)
see also mechanism method
critique, 163–4, 165–6
plasticity theorems, 140–2, 144–7, 163–4, 165–6, 182–3

L

laboratory tests, 219–20
Lagrange multipliers, 182–3, 186–7
lateral restraints, 226–30
lateral-torsional buckling, 77–8, 197–8, 219–31
Liège, Belgium, 164
limit analysis
concepts, 139–41, 163–93
constrained optimisation problems, 164–93
linear programming, 163–93
LINPROG, 184–6
MATLAB, 183–9
theorems, 139–41, 163–6
limit states design (LSD)
concepts, 74–5, 219–30
serviceability limit state, 75, 219–30
ultimate limit state, 75, 223–30
limitations of plastic design method, 76–9
linear elastic analysis
see also elastic. . .; incremental elastoplastic analysis
computer programs, 1, 108–16
concepts, 1–2, 72, 73–4, 81–3, 108–16, 197–200, 219
plastic designs, 73–4, 197–200, 219
tools, 1–2
linear interpolation, design actions, 117–18, 155, 212
linear programming (LP)
concepts, 164–93
discrete plane frame problem, 164, 175–93
dual statements of the discrete limit analysis problem, 182–3
limit analysis, 163–93
LINPROG, 184–6
MATLAB, 164, 175, 183–9
optimal plastic design problem, 164, 189–93
piecewise linear yield conditions, 178–82
pinned gable frame example, 172–5
simple propped cantilever example, 167–72, 184–9
spreadsheet applications, 164, 166–75, 183–92
linear/non-linear contrasts, yield surface, 104–5
LINPROG, 184–6
load and resistance factor design (LRFD)
see also limit states design
concepts, 74–6
load transformation
see also coordinates
concepts, 11–13, 28–32
load vector, concepts, 18–19, 20, 21–7, 28–32, 44–5, 81–3

load–deflection curves, 58–9, 72–3, 100–6, 107–37
loads, 18–19, 20, 21–7, 28–32, 44–50, 55–69, 72–3, 74–5, 81–3, 90–2, 100–6, 107–37, 151–5, 206–12, 216, 219–30
  cyclic loading, 219–21
  elastic–plastic behavior, 55–69, 81–3
  LSD, 74–5, 219–30
  point loads, 27–32, 126–30
  second-order effects, 76, 213–16
  slope-deflection equations, 10–11, 58–69, 73
  thermal loading, 45–50, 206–12, 216
  unloading, 58–69, 90–2, 124–5, 151–5
local buckling effects, plastic designs, 76–9, 197–8, 213–14, 219–31
local coordinate system, concepts, 8–9, 11–14
lower bound theorem *see* static theorem...
LP *see* linear programming
LP solvers, 167–89
LRFD *see* load and resistance factor design
LRFD–ASD national standard, 76
LSD *see* limit states design

M
manual methods of plastic analysis
  concepts, 1, 76, 139–62, 163–4, 175
  critique, 163–4, 165–6
  problems, 159–62
material types, ductility, 195–6
mathematical programming
  see also linear programming
  concepts, 55, 164–93
  historical background, 164
MATLAB, 164, 175, 183–9
matrix formulations
  see also flexibility...; stiffness...
  concepts, 2, 55–80
matrix stiffness, 2–3, 6–9
  see also direct stiffness method
maximum moment capacity, 58–9
MEA *see* modified end actions
mechanism method
  see also kinematic theorem...
  concepts, 139, 141–2, 144–7, 155–62, 163–4
member forces
  calculation, 20–7, 28–32, 43–5, 53, 155–7
  collapse calculations, 155–7
  concepts, 2, 9–10, 11–14, 20–7, 28–32, 43–5, 53
  diagram sign convention, 22–7
  flexibility method, 2, 142–4
member slenderness, design issues, 228–30
member stiffness matrix
  concepts, 9–11, 13–18, 20–7, 32–45, 83, 92, 198–200, 203–6
  condensation method, 32–45, 198–200
  derivation, 9–11
  global coordinate system, 13–14, 15–18, 30–2, 35–45
  pins, 32–45, 99–100
Merchant–Rankine formula, 213
methods of solution, 19–20
Microsoft Excel, 66, 112–16, 164, 167–75, 183–92, 209
  see also spreadsheet applications
modified end actions (MEA)
  see also modified fixed-end force vector
  concepts, 9–11, 35–6, 40–1, 101–6, 203–6
  definition, 101, 203
  foundation settlements, 203–6
modified fixed-end force vector, 35–6, 40–1, 101–2, 203–6
  see also modified end actions
  pins, 35–6, 40–1, 101–2
modulus of elasticity, 46–50, 53, 208–12, 227–30
moment amplification factor, 76
moment–curvature relationship
  concepts, 59–72, 78–9, 227–30
  elastic behavior, 59–62, 227–30
  elastic–plastic behavior, 59–72, 227–30
  fully plastic section, 69–73, 78–9, 107–37
  numerical method, 65–9

N
NATO, 164
'natural' stress resultants, 176–83
normality rule, concepts, 91–2, 175–83
numerical methods
  concepts, 19–20
  moment–curvature relationship, 65–9

O
optimal plastic design problem, 164, 189–93
optimisation issues, 164–93

P

partial collapse
  see also collapse
  continuous beams, 147–51
  definition, 148
  portal frames, 147, 151–5, 219–20
piecewise linear yield conditions, concepts, 178–82
pins, 4–6, 23–7, 32–45, 51–3, 82–3, 99–102, 136–7, 161–2, 172–5
  see also frames
  concepts, 4–6, 32–45, 82–3, 99–102, 172–5
  condensation method, 32–45, 51
  elastoplastic analysis, 32, 82–3, 99–102
  extrafreedom method, 32, 41–5, 51
  freedom codes, 32–45
  model methods, 41–5, 82–3
  modified fixed-end force vector, 35–6, 40–1, 101–2
  rotations, 32–45, 51–3
  statical indeterminacy, 4–6
  stiffness method, 4–6, 23–7, 32–45, 99–100
  treatment, 32–45
plastic analysis
  see also elastoplastic analysis
  computer programs, 1, 76, 108–16, 164, 166–75, 183–9
  concepts, 1–2, 55–80, 116–18, 139–62, 195, 201–3
  critique, 195
  definition, 3
  distributed loads, 126–30, 144–7
  elastic approach, 1, 55–69
  elastic–plastic behavior, 55–69, 72–4, 92–106, 196–8
  foundation settlements, 201–3
  linear programming, 164–93
  manual methods, 1, 76, 139–62, 163–4, 175
  stress–strain relations, 9–10, 56–69, 77–8, 139–40, 196–8
  uses, 3, 116–17, 139–62, 195
plastic behavior of structures
  concepts, 55–80, 139–62
  problems, 78–9
plastic designs
  see also design...
  benefits, 74
  computer programs, 1, 76
  concepts, 1, 46, 73–9, 86–7, 163–93, 197–8, 207–16, 219–31
  definition, 73
  design codes, 46, 75–6, 86–7, 197–8, 207–16, 219–31
  ductility, 75–6, 77–9, 195–217, 219, 223–30
  elastic designs, 73–4, 76, 197–200, 219–20
  lateral-torsional buckling, 77–8, 197–8, 213–14, 219–31
  limitations, 76–9
  linear elastic comparison, 73–4, 197–200, 219
  LSD, 74–5, 219–30
  optimal plastic design problem, 164, 189–93
  statical indeterminacy, 74
  ultimate limit state, 75, 223–30
plastic flow theory
  concepts, 55–9, 81–106
  dissipative mechanical energy, 91–2
  stress states, 91
  yield surface, 89–92
plastic hinges
  see also plastic moment
  concepts, 3–4, 68–9, 72–3, 76–9, 81–106, 107–37, 139–44, 165–93, 195–217, 224–30
  incremental elastoplastic analysis, 107–37, 141
  stability issues, 227–8
  unloading, 124–5, 151–5
plastic moment
  see also plastic hinges; shape factors
  axial forces, 83–7, 119–24
  concepts, 57–69, 72–3, 76–9, 100–6, 108–16, 119–24, 142–3, 160–2, 195–6
  general expression, 64–5
  reduced capacity due to force interaction, 83–7, 119–24, 157–9
  shear forces, 83–7, 119–24, 219–31
plastic multiplier, concepts, 91–2, 102–3, 175
plastic rotation, 72–3, 75–6, 77–9, 81–3, 91–2, 141–2, 150–1, 195–217, 219–30
plastic rotation capacity
  see also ductility
  concepts, 195–217, 224–30
  definition, 195–6
  determining factors, 196
  steel, 196–200

plastic rotation demand
  concepts, 196–200, 215–16
  condensation method, 198–200
  definition, 196
  examples, 199–200
plastic section modulus, 69–72, 79, 227–30
plasticity
  concepts, 139–62, 163–93
  kinematic theorem (upper bound theorem) 140–2, 144–7, 163–4, 165–6, 182–3
  static theorem (lower bound theorem) 86–7, 140–1, 144–7, 155–7, 163–93
  theorems, 139–41, 163–6
  uniqueness theorem, 141
plastification, concepts, 57–69
point loads, 27–32, 126–30
  see also internal loads
portal frames, 1–2, 113–18, 128–30, 151–62, 209–10, 219–20
positive bending
  see also bending...
  diagram convention, 22–7
positive shearing
  see also shear...
  diagram convention, 22–7
primary structure, flexibility method, 2, 142–4
proportional loading, 107–37
propped cantilever beams, 34–5, 37–8, 109–12, 130–1, 149–51, 159–62, 167–75, 184–9, 202–3, 208–12, 221–3
pure bending, 55, 61–2, 69–72, 82–3, 87, 91–2, 96–106, 108–16, 119–24, 131–7, 175–93, 204–6, 228–30
  axial–bending interaction, 99–106, 119–24, 175–93, 197–200, 205–6, 228–30
  successive approximation method, 107, 119, 122–4, 157–9

R

redundant forces, concepts, 2–3
reserve strength
  see also plastic designs
  concepts, 73–9
rigid joints, statical indeterminacy, 4–6
rigid plastic analysis
  see also plastic analysis
  concepts, 139–62, 223
roller joints, statical indeterminacy, 4–6, 30–2
rotations, pins, 32–45, 51–3

S

sagging beams
  see also positive bending
diagram convention, 22–7
second-order effects, 76, 213–16
self-equilibrated elements, 177–83, 206–12, 221–3
serviceability limit state
  concepts, 75, 219–30
  design issues, 219–30
  LSD concepts, 75, 219–30
settlement effects *see* foundation settlements
shape factors
  common cross sections, 64–5
  concepts, 58–69
  definition, 58
shear forces
  see also design actions
  concepts, 10–11, 22–32, 51–3, 82–7, 92, 116–17, 157–9, 219–31
  diagram sign convention, 22–7
  reduced plastic moment capacity, 83–7, 119–24, 157–9
sign convention, member force diagrams, 22–7
slenderness
  concepts, 224–30
  definition, 224
slope-deflection equations, 10–11, 58–69, 73
Smith, Lloyd, 183
soil–structure interaction problems, 200–1
  see also foundation settlements
spreadsheet applications, 1–2, 66–9, 107, 108–16, 164, 166–92, 209
  see also computer programs; Microsoft Excel
spring systems, stiffness method, 6–8
static theorem (lower bound theorem)
  critique, 163–4, 165–6
  plasticity theorems, 86–7, 140–1, 144–7, 155–7, 163–93
statical indeterminacy
  concepts, 2–9, 20, 30–2, 50–1, 74, 143–4, 148
  degrees, 2, 4–6, 30–2, 50, 143–4, 148, 153–5
  direct stiffness method, 3, 6–9
  frames, 4–6, 30–2
  importance, 3–4
  joints, 4–6, 30–2

plastic designs, 74
statical method *see* static theorem
steel
class of sections, 224–6
ductility, 196–200, 223–30
elastic–plastic behavior, 55–69, 196–8
plastic rotation capacity, 196–200
temperature effects, 45–50, 206–12
stiffness method
concepts, 1–54, 55–80
elastoplastic analysis, 2–3, 32
elastoplastic stiffness matrix, 91–106, 119–24
internal loads, 2, 27–32, 33–45, 126–30
member stiffness matrix, 9–11, 13–14, 15–18, 20–7, 83, 92, 198–200, 203–6
methods of solution, 19–20
pins, 4–6, 23–7, 32–45, 99–100
plastic behavior, 55–80
problems, 50–3
structural analysis, 2–54
structure stiffness matrix, 7–9, 14–20, 21–7, 31–2, 83, 92, 113–16, 119–24
temperature effects, 45–50, 53
usage, 2–3
strain hardening
design issues, 57, 73–4, 197–200
elastic–plastic behavior, 56–69, 73–4, 197–200
stress–strain relations
concepts, 9–10, 56–69, 77–8, 139–40, 196–8
elastic–plastic behavior, 56–69, 77–8, 81–3, 139–40, 196–8
structural analysis
computational issues, 1–2
computer programs, 1–3, 66–9, 76, 107, 108–16, 119–20, 122–3, 164, 166–75, 183–9
concepts, 1–54
degrees of freedom, 3–9, 28–32, 192–3
direct stiffness method, 3, 6–9
elastoplastic stiffness matrix, 91–106, 119–24
flexibility method, 2, 142–4
internal loads, 27–32, 33–50, 126–30
matrix formulations, 2
member stiffness matrix, 9–11, 13–14, 15–18, 20–7, 83, 92, 198–200, 203–6
methods of solution, 19–20
pins, 4–6, 23–7, 32–45, 82–3, 99–100
problems, 50–3
second-order effects, 76, 213–16
statical indeterminacy, 2–9, 20, 30–2, 50–1, 74, 148
stiffness method, 2–54
temperature effects, 45–50, 53, 75, 206–12
structural engineering design offices, 111
structure stiffness matrix
assembly, 14–18, 21–7, 31–45
compatibility condition, 14–16
concepts, 7–9, 14–20, 21–7, 31–45, 83, 92, 113–16, 119–24
equilibrium condition, 14, 16–18, 20–7
methods of solution, 19–20
pins, 32–45, 99–100
successive approximation method
concepts, 107, 119, 122–4, 157–9
force interaction effects on plastic collapse, 119, 122–4, 157–9
sway mechanisms, partial/complete collapse, 152–9, 210–11

T

temperature effects
coefficient of linear expansion, 46–50, 208–12
concepts, 45–50, 53, 75, 195, 206–12
cross-sectional area, 46–50, 206–12
failure temperatures, 195, 206–12
internal loads, 27, 45–50, 206–12
modulus of elasticity, 46–50, 53, 208–12
temperature gradient, 47–50
uniform temperature, 45–7, 207–11
temperature gradient, concepts, 47–50
thermal loading, 45–50, 206–12, 216
see also temperature effects
Tin-Loi, F. 163–93
torsion, 77–8, 82–3, 197–8, 213–14, 219–31
torsional restraints, 226–7
transformations
see also load...
concepts, 11–13, 28–32, 39–45
coordinates, 11–13, 28–32
displacement transformation, 12–13
load transformation, 11–13, 28–32
transformed section method, temperature effects, 48–50
truss members, stiffness matrix, 39–40

U
UDL *see* uniform distributed loads
ultimate limit state, LSD concepts, 75, 223–30
ultimate settlement factor, concepts, 202–3
unbraced length, design issues, 226–7
uniform distributed loads (UDL)
  concepts, 144–7, 149–51, 160
  continuous beam end span, 149–51, 160
  external work calculation method, 146–7
uniform temperature
  see also temperature effects
  concepts, 45–7, 207–11
uniqueness theorem, plasticity theorems, 141
United States
  see also AISC
  design codes, 74–6, 215, 219, 224–31
unloading
  concepts, 58–69, 90–2, 124–5, 151–5
  plastic hinges, 124–5, 151–5
upper bound theorem *see* kinematic theorem...

V
von Mises yield criterion, 85–6

W
Waterloo conference (1977) 164
web slenderness, design issues, 228–30
wind loads, 219–20

Y
yield conditions, piecewise linear yield conditions, 178–82
yield curvature, concepts, 62–9
yield function
  concepts, 89–92
  definition, 89
yield moment
  see also shape factors
  concepts, 57–69
yield surface
  see also axial forces; bending moments
  concepts, 83, 87–99, 104–5, 119–24, 181–3
  definition, 87
  elastoplastic analysis, 92–9, 104–5, 119–24
  linear/non-linear contrasts, 104–5
  plastic flow rule, 89–92
  stress states, 91
yielding, pure bending, 55, 61–3, 69–72, 82–3, 87, 91–2, 96–106, 108–16, 119–24, 131–7, 175–93, 204–6, 228–30
Young's modulus, 10

Made in the USA
Middletown, DE
12 November 2026

36932797R00144